9/24/80

D0983176

LIFE BEYOND EARTH

*The Intelligent
Earthling's Guide to
Life in the Universe*

LIFE
BEYOND
EARTH

The Intelligent

Earthling's Guide to

Life in the Universe

by
Gerald Feinberg
and Robert Shapiro

WILLIAM MORROW AND COMPANY, INC.
New York   1980

Library of Congress Cataloging in Publication Data

Feinberg, Gerald, 1933-
    Life beyond Earth.

    Bibliography: p.
    Includes index.
    1. Life on other planets.   I. Shapiro, Robert,
1935-   joint author.   II. Title.
QB54.F34   574.999   80-11832
ISBN 0-688-03642-2
ISBN 0-688-08642-X (pbk.)

Printed in the United States of America

First Edition

1   2   3   4   5   6   7   8   9   10

BOOK DESIGN BY MICHAEL MAUCERI

*To Sandra and Michael; and to Barbara,
Jeremy and Douglas*

with love

Preface

Our book is concerned with the nature of life, its possible forms, and its distribution throughout the Universe. There has been a great deal of interest in the subject of *intelligent* extraterrestrial life and several books in past years have dealt with it. We feel that the existence of life beyond Earth, apart from being a prerequisite for the existence of intelligent life, is in itself a subject of great interest. The life habits of even the most humble Earth creatures have fascinated biologists and naturalists since ancient times. If the homes of life are extended to include much of the Universe, the importance of the topic will be expanded accordingly.

The subject of extraterrestrial life by its very nature cuts across scientific disciplines. Its name suggests a combination of astronomy and biology, two sciences that in the past have been dominated by the results of experimentation instead of the sweeping theoretical structures of sciences like physics. Since there exists little or no data concerning extraterrestrial life, many experimental scientists have been indifferent to the subject. Others have assumed that, if it exists at all, it will be somewhat like life on Earth, and they have often used the term "life as we know it."

In this book we, a physicist and a biochemist, wish to speculate on life as we do not know it. It is not our intent to add to the growing volume of science fiction literature, but to see what can be deduced theoretically from the known laws of science. We have used what is known about life on Earth to infer essential features that any form of life must possess. We then investigate how these general features might express themselves in the specific environments that are found in various parts of the Universe. In other words, we have gone from Earthlife itself to a general definition of life. With this general definition, and a knowledge of the universal laws of physics and chemistry, we then try to imagine what specific types of life can develop in each environment.

The result of this process is different in some respects from most scientific writing. While much of our book is a straightforward exposition and analysis of known science and its consequences, there are sections, especially in Chapters 8, 12, and 13, where we have been more speculative. We say this not in apology, but rather to indicate to our readers that we are well aware of the different status of these parts of the book.

In order to carry out this program, even in outline, we have had to tie together many strands from different areas of science. The unsupported knowledge of any two scientists is insufficient to do this. Therefore, we have called on many other scientists and scholars for help. Their assistance has involved instruction in specific areas outside of either of our specializations, descriptions of some of their views on extraterrestrial life, and comments on some of our conclusions. We are profoundly grateful for this help, without which it would have been difficult or impossible to write this book. Those whom we wish to thank include: Arthur Bain, John Brockman, Erwin Chargaff, Russell Doolittle, Henry Foley, E. Imre Friedmann, Michael Hart, Norman Horowitz, Harold P. Klein, James Lawless, Joshua Lederberg, Martin Leibowitz, Richard Lemmon, Gilbert Levin, Willard Libby, Joaquin Luttinger, Robert MacElroy, Katinka Matson, Stanley Miller, Sidney Morgenbesser, Harold Morowitz, Randall Murphy, Ernest Nagel, Robert Novick, Christopher Norwood, Leslie Orgel, Tobias Owen, Vance Oyama, Malvin Ruderman, Barbara Sakitt, J. William Schopf, Harold Schuckman, Paul and June Shepard, Gerald Soffen, Nelson Spencer, Guenther Stotzky, Benson Sundheim, Menasha Tausner, Pat Thaddeus, and Richard S. Young.

We also thank Jeremy Bernstein for reading the manuscript and for his helpful suggestions, our typists, Gena Vandestienne, Anne Billups, Jill Gilmore, Betty Martin, and Irene Tramm, for their assistance, Ed Hamza for photographing a number of plates, Fred Golden for help in obtaining a photograph, and Bob Muller for his contribution in Figures 1, 2, and 32.

Finally, we wish especially to thank our editor, Maria Guarnaschelli, for her continuing enthusiasm and encouragement, and her valuable suggestions.

Contents

List of Figures

List of Plates

Credits

Drawing: Start of Section I, by Daniel Scarola.

Photograph: Start of Section II. This photograph was taken from an article, "Rotating Chemical Reactions," by A. T. Winfree in the June 1974 *Scientific American*. It is reprinted by permission of the photographer, Mr. F. W. Goro. © by Fritz Goro, 1974.

Cartoon: Start of Section III. Drawing by R. Grossman; © 1962. The New Yorker Magazine, Inc.

Poem: Start of Chapter 8. © 1946 Saturday Evening Post, © Robert A. Heinlein, 1974.

Plates 1, 2: Courtesy of the American Museum of Natural History.

Plate 3: Courtesy of Monroe Yoder, Department of Biology, New York University.

Plate 4: Photographed by Mr. Ed Hamza.

Plate 5: "Biochemical Pathways," by Gerhard Michal, revised 1974, Copyright Boehringer Mannheim GmbH.

Plate 6: O. L. Miller, Jr., B. A. Hankalo, and C. A. Thomas, Jr., *Science, 169*, 392–395, July 24, 1970. Copyright 1970 by the American Association for the Advancement of Science.

Plate 7: A. H. Knoll and I. S. Barghoorn, *Science 198*, 396–398, October 26, 1977. Copyright 1977 by the American Association for the Advancement of Science.

Plate 8: Percival Lowell, "Mars as the Abode of Life," The Macmillan Company, New York, 1908.

Plates 9, 10, 11, 12, 13, 14, 15, 16, 17: Courtesy of NASA.

Plate 18: Reprinted from the *Astronomical Journal 82*, 249, 1977; courtesy of Dr. P. J. E. Peebles and of the *Astronomical Journal*.

Plate 19: Courtesy of the Hale Observatories.

Plate 20: Betelgeuse. Courtesy of the Kitt Peale National Observatory.

Plate 21: Courtesy of the Hale Observatories.

Figures 28, 29, 30: Reprinted, with slight modification, from *Physics of Dense Matter*, ed. C. Hansen, pp. 127, 129, Dordrecht, Holland, D. Reidel, 1974, courtesy of Dr. M. Ruderma and of D. Reidel Publishing Co.

"The properties of Earthlife are written in the structure of DNA."

—Daniel Scarola

"Spirals that appear spontaneously in certain chemical reactions indicate how order can arise in a nonbiological system." —F. Goro

Authors' Note About Units and Notation

In this book we will use exponential notation to express very large and small numbers. With this system we can avoid cluttering up the page with many zeroes. For example, the number 1,000,000,000,000, which in words is a thousand billion or a million million, in the exponential system is simply 10^{12}. To convert an exponential number to a conventional one, write down "1," followed by the number of zeroes in the exponent (twelve in the above case). Numbers with only a few zeroes can also be indicated in exponential notation (though we will not do so in this book). One can be written 10^0, ten as 10^1, a thousand as 10^3, a million as 10^6, and a billion as 10^9.

What does a negative exponent indicate? When we write 10^{-3}, we mean one divided by 10^3, which we can otherwise indicate as $1/10^3$, $1/1000$, the words one thousandth, or the decimal 0.001. Similarly, 10^{-9} means the same as $1/10^9$, $1/1,000,000,000$, one one-billionth, or 0.000000001.

We use the metric system for units of length, mass, and temperature. The unit of length is the meter, which is about thirty-nine inches, or roughly 10 percent more than a yard. Some other lengths commonly used are the centimeter (one-hundredth of a meter, 0.39 inches) and the kilometer (one thousand meters, 0.62 mile).

The unit of mass is one gram. There are about twenty-eight grams in an ounce. A kilogram, one thousand grams, is about two and a fifth pounds. The temperature scale is centigrade. The following table gives the relation between several centigrade (C) and Fahrenheit (F) temperatures:

Degrees C	Degrees F	Degrees C	Degrees F
−200	−328	40	104
−100	−148	100	212
−50	−58	200	392
0	32	500	932
10	50	1,000	1,832
20	68	2,000	3,632

To convert a centigrade temperature to Fahrenheit, multiply by nine, then divide by five, and add thirty-two. We list very low temperatures in the Kelvin scale as well as in centigrade. To convert from Kelvin to centigrade, simply add 273 degrees.

Chapter 1

Life in the Universe

Life prevails on the planet Earth. Sometimes this is immediately and dramatically visible, as in the Amazon jungle, the North American forests, and a drop of pond water seen under the microscope. In other places on Earth, such as hot deserts and the dry, cold valleys of Antarctica, the presence of living beings is less obvious. But even there, a careful observation usually reveals that there are inhabitants occupying special niches to which they are well adapted. Through a history of almost four billion years, life has captured most of the Earth's surface and transformed it into an environment quite altered from the original one.

The situation on Moon appears to be entirely different. A number of manned expeditions to its surface have discovered no form or substance that can be identified with life. This stark contrast in the condition of the two worlds that have been explored raises a question about the remainder of the Universe: Are there many other places which are, like Earth, richly endowed with life, or is our planet a unique oasis in a vast, utterly barren cosmos?

This book is dedicated to that question. In it we present a picture of a fertile Universe, teeming with diverse life forms: The generation of life is a common event in the Universe, a preferred form of self-expression of matter. Life exists not only on the solid surfaces of planets but also in dense planetary atmospheres, in rarefied interstellar clouds, and even inside of some stars. In various environments, very different life forms evolve, in harmony with the conditions present there. Some of these life forms would be familiar to us, but some would be so alien that we would have difficulty in recognizing them as alive. While a number of extraterrestrial life forms could be even more primitive than bacteria, others might be quite advanced, with intelligence, other physiological capabilities, and technological achievements far exceeding our own.

Our analysis is based on late twentieth century science, with its dramatic advances in understanding of the genetic mechanisms of life on Earth, the nature of the solar system, and the history of the Universe. Despite these advances, there are still many questions relevant to the scope of life in the Universe that cannot be answered yet: the way in which life originated on Earth, the detailed nature of other planets and satellites in our solar system, the extent of planets elsewhere in our galaxy. We have tried to fill these gaps with extrapolation from current knowledge and with imagination. Others have started with the same basis of fact and arrived at different pictures of the Universe. We hope that the contrast produced by these alternative visions will stimulate interest and speed the day when the issue will be resolved.

THE DEBATE ON EXTRATERRESTRIAL LIFE

The question of the extent of life in the Universe is of course not new, but rather, has been debated for centuries. At least one martyr, Giordano Bruno, was burned to death for his views on the subject. Although this topic has had no direct effect on the daily affairs of humanity thus far, it nonetheless has the power to provoke strong feelings. For it affects the framework of values through which we perceive the purpose of our own lives, the goals of humanity, and the place of mankind in the Universe. Arguments in the past have often been based more on religion and philosophy than on science. Yet it is important that we review them here because remnants of this thinking still recur in many supposedly scientific discussions today.

An important theme in Western religion has been the special relation of the human race to the Deity who, according to religious traditions, created it. In ancient and medieval times the unique relation of man to God was emphasized by astronomical schemes that placed the Earth at the center of the Universe, with the various celestial bodies revolving about it. This support to our vanity was removed by the findings of modern astronomy. The Earth was relegated to the position of a minor planet circling an undistinguished star, which itself is far from the center of its galaxy. The galaxy is also just one of a large number of similar bodies which dot the Universe.

This humbling effect of our insignificant location would be counteracted, and our importance restored, if Earth were the only site of life in the Universe. The contents of this planet would be unique and therefore precious. We would be alone in the Universe with God, our responsibility would be vast, and our actions would truly have cosmic significance.

This point of view was expounded more than a century ago by an English educator and scientist, Dr. William Whewell, in a debate conducted in print with a Scottish physicist, Sir David Brewster. Although each proponent made the best use that he could of the science available at the time, the central struggle was conducted in religious terms. A key argument put forward by Brewster in behalf of a populated Universe concerned the purpose of God in creating such a vast domain. It was unthinkable that He would have created multiple worlds unless He had intended them as homes for life. The answer to this argument, throughout history, has been that the purposes of God need not be understood by us. Today, of course, many scientists, ourselves included, would answer instead that the properties of the Universe are governed by its physical laws and not by the motives, scrutable or inscrutable, of a Deity. If one wishes, however, to return to the Whewell-Brewster debate in the spirit in which it was conducted, a new modern reply can yet be formulated: The vast Universe was created to be inhabited, not by alien life forms, but by us. Our cosmic purpose is to spread the spark of life and consciousness to the far corners of creation. There is a great deal of romantic appeal in this position, which may explain why many have argued up to the present for a Universe in which Earth is the only home for life.

The opposing viewpoint, that of a Universe full of life, also has a long history, going back at least to Cusanus in the fifteenth century. It is one aspect of what the cultural historian Arthur Lovejoy has called the principle of plenitude: that everything possible is realized in nature. The view of a Universe teeming with life has its own set of attractive features, which have led many writers to argue for it in speculative papers or to invoke it in literary works.

The fiction writers who have speculated about alien life forms may further be divided into two groups—those who described benign types of life and those who portrayed hostile ones. In making a choice, the author usually is projecting his hopes or fears about the basic nature of humanity, rather than making a rational guess about that of extraterrestrial life. In his famed novel *War of the Worlds* (1898), H. G. Wells described the invasion of England by intelligent, belligerent, tentacled Martian beings. He commented:

> *And before we judge of them too harshly we must remember what ruthless and utter destruction our own species has wrought, not only upon animals, such as the vanished bison and the dodo, but upon its inferior races. The Tasmanians, in spite of their human likeness, were entirely swept out of existence in a war of extermination waged by European immigrants, in the space of fifty years. Are we such apostles of mercy as to complain if the Martians warred in the same spirit?*

The literature and films of science fiction have been filled with alien creatures come to punish us. The intelligent ones have mimicked the savage parts of our own nature. A variety of others have also been described, from ferocious monsters to lethal plagues. These have lacked the capacity to act out of their own vengeful purposes, but presumably were the agents of a wrathful Deity, somewhat like the plagues described in the Bible. In the alternative view, our species has been portrayed as well-intentioned but clumsy children badly in need of advice and a pat on the back from our older and wiser extraterrestrial brothers. This yearning for help and companionship was well illustrated in the film *Close Encounters of the Third Kind*. An earlier example of this theme can be found in the book *Life in the Stars* by Sir Francis Younghusband (1927), in which the inhabitants of other worlds were described as higher beings with angelic qualities. Their superior status derived from their greater spirituality; they also possessed sensory organs whose quality and delicacy exceeded ours.

Most arguments put forward in the past about life in the Universe tell us more about the values and psychology of the debater than about the subject under debate. The relevant scientific data has been insufficient to support the conclusions drawn. For example, let us return to the Whewell-Brewster debate of the 1850's. A considerable amount of astronomical knowledge was available, but it was not sufficiently detailed. (Thus, disputes occurred over whether there was rain on Moon, and whether the interior of the Sun might be cool rather than hot.) Moreover, there was complete ignorance of the basic biochemical mechanisms that support our life. Lacking this knowledge, one could only treat life as a universal principle of nature and assume that in other parts of the Universe, it would be similar to life on Earth. It was appreciated that light, air, water, and a certain temperature range were vital to terrestrial life. The crux of the scientific debate then hung on matters of climatology: Were these necessities present in sufficient amounts on Jupiter, the Sun, or Moon?

Early in this century, similar considerations arose in a dispute over intelligent life on Mars. The astronomer Percival Lowell, in supporting the existence of civilizations on Mars, estimated the surface temperature of that planet as 10°C. His critic, the evolutionist Alfred R. Wallace, correctly deduced that it was much colder.

The modern scientific viewpoint holds that the Universe was not created for our benefit. If extraterrestrial life exists, it does so in accord with the laws of nature. It has not been put there in order to punish or reward us, or to fulfill any other divine purpose. We can best determine what may exist if we set aside our hopes and fantasies, and use our powers of observation and reason instead.

INFERENCES FROM EARTHLIFE

One basis for reasoning about life in the Universe is the knowledge that we have about life on Earth (which we will call Earthlife). As this is the only type of life that we have observed, it is an essential source of information. But we must be cautious in generalizing too quickly about Earthlife, lest we fall into the error of one of the blind men describing an elephant.

A remarkable modern discovery has been that of the biochemical unity of the diverse living creatures of Earth. Ants, plants, and elephants all employ two particular substances called proteins and nucleic acids in vital roles. The chemicals which they use for many other purposes are also generally the same. The complete set of chemicals utilized by Earthlife, however, represent only a very small fraction of the vast number that may be produced by natural processes. We believe that these particular substances are present in life here not because they alone, of all possible materials in the Universe, can serve as a basis for life, but because of the specific local environment and historical circumstances on this planet. In order to appreciate the full role of life elsewhere in the Universe, we must extend our imaginations beyond the chemical, temperature, and environmental boundaries that characterize Earthlife and consider a wider range of possibilities.

THE CARBAQUISTS AND OTHER PESSIMISTS

A number of scientists, some quite prominent, take a much narrower view of the prospects for life elsewhere than on Earth. They start with the biochemical unity of Earthlife and conclude that extraterrestrial life must share this unity and be similar to life here. In the present state of science, the issue cannot be definitely settled, and these scientists are certainly entitled to their opinions. Unfortunately, such opinions are often expressed in print as if they were fact and thus become self-fulfilling prophecies. Experiments are designed, as in the Viking investigations on Mars, that would detect Earthlife but miss alternative life forms. Negative results can then be represented as indicating not just the absence of Earthlife, but of all life.

This point of view has been accepted as fact not only in the space program but also in museum exhibits, textbooks, and other public media. We will therefore spend some time in this book in considering the scientific basis for it. We have coined the term "carbaquist" for those who believe that extraterrestrial life must resemble Earthlife. Carbaquists believe that all life must be based on the chemistry of *carbon* compounds, and must operate in an *aqueous* (water) medium. Some extreme carbaquists argue that the very same vital chemicals, such as nucleic acids and proteins, must occur in extraterrestrial life as in Earthlife. We quote George Wald, a Harvard biochemist who won the Nobel Prize (for his studies on the chemistry of vision, not his views on extraterrestrial life): ". . . So I tell my students: learn your biochemistry here and you will be able to pass examinations on Arcturus."

While many carbaquists are willing to consider the possibility of life on other Earthlike planets, others proceed to restrict the scope of life elsewhere even further. It is possible that even a small change in the circumstances of a planet from those of Earth would lead to a great difference in the overall environment. For example, the astronomer Michael Hart has argued that if Earth were a slightly different distance from the Sun, or if the size or brightness of the Sun were changed slightly, environmental conditions would be extremely different than those of the Earth we know, and the type of life we have here could not have evolved on such a world. So if Earthlife is the only possible type of life, very little will exist elsewhere, for there will be few planets which are in just the right position relative to precisely the right star to have an environment such as we have on Earth. On the other hand, if life can develop to fit many different environments, as we believe, then this argument would only imply that most life elsewhere is very different from Earthlife.

Carbaquism, together with Hart's arguments on planetary evolution, would still allow the existence in the Universe of a few other worlds bearing life like ours. Even this possibility would be eliminated by some views about the unknown steps in the origin of Earthlife. While scientists have demonstrated how some simple chemicals necessary to Earthlife may have been synthesized by random processes during the early history of Earth, it is not understood how a mixture of their simple chemicals was transformed into a more complex organized state capable of further evolution. We do not know whether that transformation was a likely event inherent in the laws of nature, or whether it represented a unique, once-in-eternity long shot. If the probability of the transformation is very small, then even the presence of life on Earth would be surprising, perhaps miraculous. Its presence elsewhere, even on worlds environmentally similar to Earth, would require repeated miracles. The effect of this line of reasoning is to remove the question of extraterrestrial life from the realm of science and return it to theology.

WHY SHOULD WE CARE ABOUT
EXTRATERRESTRIAL LIFE?

In later chapters we will deal with these pessimistic arguments concerning extraterrestrial life, describe possible alternate forms of life, and consider various strategies for detecting them. Before we do so, there is a more fundamental question to consider: Why search at all? There are now, as always, many pressing problems close at hand. Why should we turn any part of our attention away from them to consider such exotic and remote questions?

One reply to the above queries involves the purposes of science itself: It is a human activity. We have a *need* to know the answers to fundamental questions about nature. What is Life? ranks with What is Matter? and What is Consciousness? in this category. It is very difficult to deduce what the general properties of a phenomenon such as life might be when one has only a single type of it to study. Imagine the difficulties of a musically naïve person in deducing the existence of Beethoven symphonies, Indian ragas, and electronic music, given only a recording of *Oklahoma!* If several alternative forms of life were available for study, we would have a true comparative biology. We could better decide which properties of matter are necessary, and which incidental, to life. We would undoubtedly learn a lot of fundamental chemistry as well. Almost all of the early discoveries about the chemistry of carbon compounds resulted from the study of materials extracted from living things on Earth. On the other hand, if an exhaustive search of a variety of cosmic environments turned up only further examples of Earthlife, it would suggest that our kind of life was the only one that had a reasonable chance of developing or even existing, given the laws of nature.

Apart from their importance to theoretical biology, extraterrestrial life forms would have their own intrinsic interest. Most of us are fascinated by the natural history of Earthlife: the social life of bees, the mating habits of birds and fish, the strategies of various species for defense, communication, and raising of the young. We identify with the struggle for survival, whether it is that of a plant growing in a crack in a rock, an insect seeking water in an arid desert, or a lichen defying the exposure of a windswept mountaintop. What additional wonders and surprises may await us as we explore the habits of life forms that have evolved and survived in environments that are very strange to us?

Life is the most interesting manifestation of the properties of matter that we know. It is inconceivable that we could have guessed its existence just from what we know of inanimate objects, even though we are equipped with the sophisticated sciences of physics and chemistry. But once we have obtained the clue from nature that something as wonderful as Earthlife can emerge from the interplay of matter and energy, we can let our minds roam further and try to anticipate what similar marvels the workings of natural law can bring forth in alien environments. We believe that most other attempts to do this have not been sufficiently imaginative in examining the possibilities for extraterrestrial life. This book, our contribution to the debate that has persisted for two millennia, is an effort to expand the imaginations of our readers, and to prepare them for some of the wonder of life that future generations will discover as they explore the Universe. At a later point we shall take a journey in our imaginations to other worlds and environments of the Universe to inspect the exotic flora and fauna.

In order to do this with some basis in science rather than fantasy, we will need to know more about what life is and what the natural laws are that govern processes in both the living and nonliving portions of the Universe. In the next chapters we will consider the construction and possible origin of the one life form we are familiar with—Earthlife.

EARTHLIFE

Complete certainty now exists among essentially all biochemists that the other characteristics of living organisms . . . will all be completely understood in terms of the coordinative interactions of large and small molecules.

—J. D. WATSON

The properties of Earthlife are written in the structure of DNA.

Chapter 2

The Architecture of Life

Earthlife occurs in a bewildering variety of shapes, sizes, colors, and behaviors. A whole science, biology, was developed to study it. Yet what is most important for our discussions of extraterrestrial life is not the variety of forms and activities of Earthlife, but rather the great similarities among living things which are hidden behind the differences. These similarities exist in the way living things are constructed, especially on the submicroscopic level, and in the physical and chemical principles through which they function.

Therefore, we will not try to summarize all of the vast amounts of information that biologists have compiled. Nevertheless, one of the most striking differences among living things is the enormous range of sizes in which they come, from microbes to men and beyond. We will begin this chapter with a brief tour of Earthlife, comparing the sizes of various living things as well as their parts. We will then describe those aspects of their structure and function that are common and essential to all the forms of life we know.

A TRIP WITH **COSMEL**

Different forms of Earthlife vary widely in size, and because most of them are much smaller than we are, there is a problem in comparing them meaningfully. It is easy for us to understand the relative size of objects that we can see and touch, such as a mosquito, an elephant, and a redwood tree. It is when we consider the smaller creatures that exist among us, and their component parts, that our problems occur. Thus we can say that an amoeba, a bacterium, and a protein molecule are all very small, but their relation in size to us and to each other is not easily apparent. In another context, when we later consider parts of the Universe, we can recognize that the planet Jupiter, the Sun, and our galaxy are all very large objects. But we cannot easily relate their size to our normal human experience.

To assist us in perceiving objects large and small during the course of this book, we will use an aid for our imaginations, which we will call the Cosmic orders of magnitude elevator, or COSMEL for short. Since COSMEL exists in our minds, we can cause it to materialize alongside us at will. We do so now and note that it resembles an ordinary elevator car except that it stands freely rather than being confined to a shaft. We push the button on the door and step inside. It is also like an elevator car inside, bare of furniture and with a panel of numbered buttons, listed in a European style. Ground level is designated G (or zero), and a column of buttons with the numbers $+1$, $+2$, $+3$, etc., extends above it. Unlike most elevators, there are also buttons indicating the presence of a number of lower levels marked -1, -2, -3, and so on. These numbers indicate something about the amount of enlargement or reduction that takes place on each level that COSMEL enables us to visit. That is, on level $+3$, one meter of apparent size corresponds to 10^3 (or one thousand) meters of actual size. On level -2, one meter of apparent size corresponds to 10^{-2} meters (one centimeter) of actual size. A guide to COSMEL levels, with a list of interesting objects to be found on some levels, is given on page 437. A final novel feature is a television screen, with several dials, above the row of buttons.

First let us try COSMEL in its expanding stages. We

begin our trip in New York's Central Park, near a small zoo. The people, other animals, and trees all look normal before we enter COSMEL. We step inside, push the button marked $+1$, and then emerge. Our whole perspective has changed. We are still standing in the park, but we seem to be ten times as tall as before. The trees look like shrubs, the elephants in the zoo are only about up to our knees, and the people just come up to our ankles. In fact, through an increase by one factor of ten in height, we become taller than any animal that has ever walked the Earth. Dinosaurs such as Brontosaurus and Diplodocus have attained greater dimensions, in the range on level zero of twenty to thirty meters, but this has been in length. Their support problems were solved in part by their massive tails, which lay on the ground.

There is a good reason why animals have been so limited in size. If COSMEL had really increased our height tenfold, we would have become aware of it at once as we crumpled to the ground. Since our bodies extend in three dimensions, our weight would have increased by approximately ten times ten times ten, or one thousand times. The ability of our legs to support us, however, which depends on their two-dimensional cross-sectional area, would only have gone up by ten times ten, or one hundred times. The strain on our legs would be ten times greater than we are accustomed to, and the consequences would be disastrous. For this reason, Gulliver's giant Brobdingnagians would have had great difficulty in standing. The same problem exists for other large living things and puts a limit on the size of land animals, restricting them to dimensions not much greater than those of human beings.

Furthermore, if we were eighteen meters tall, we would have other problems. The rate at which heat escapes our body varies with the area of body surface, which would have increased by one hundred times. However, the heat we produce in our metabolism would increase as our body weight, or one thousand times. We would quickly broil in our own waste heat. Large animals can avoid this either by having a lower metabolic rate, or by developing a skin with many folds so as to increase their effective surface area.

This type of reasoning, which indicates that objects cannot always be scaled up or down in size without great changes in their structure, is sometimes called the square-cube law. It was introduced into science by Galileo in the seventeenth century. We shall have several occasions to use it in this book, as it determines a number of the properties of living things.

We have not suffered the effects described above because COSMEL has not actually changed our size. It has neither expanded nor reduced us, but has simply transported us to a model containing elaborate and extensive three-dimensional images of the objects we wish to view. These images, somewhat like holographs, can readily be expanded or reduced in size without running into trouble with the square-cube law. Also, we can highlight or suppress in the model whatever aspects of the object are uninteresting for our purposes. Finally, our model in most cases will remain static and not change in time. In a few cases, however, we will let the objects in the model move at a rate that is convenient for us to observe, as is done in some types of slow-motion or speeded-up photography. In this chapter our models will be used to show how smaller and smaller parts of living things appear, and in later chapters, to indicate astronomical objects. But we will continue to write as if we were visiting these different levels of the actual Universe, without worrying about the fact that we are not changing at all but only looking at models of the real thing.

We shall not tarry further on level +1, for there are too many wonders awaiting us. We summon and enter COSMEL, and push the button marked +2. When we step out, our horizons have broadened once more. We can readily see across Central Park and, over the multistory buildings that line its west side, to the cliffs of New Jersey. Looking down Fifth Avenue, we can see the Empire State Building. It is double our height, four meters tall, and about twenty-five meters away. We could stroll over there, if we chose, in about twenty seconds. Instead we cross the park to the American Museum of Natural History. This building is ordinarily composed of very solid stone, but in our model on level +2, we have arranged for it and most of its contents to be quite transparent. The museum's reconstructions of dinosaurs, the largest animals ever to walk on land, appear to be the size of mice on this level. In the Hall of the Sea on the ground floor, we can examine the spectacular life-size replica of a blue whale (Plate 1). This whale, whose length

Plate 1. A model of the blue whale, the largest animal on Earth. The whale would appear about this size on the +2 level of COSMEL.

on level zero can exceed thirty meters, is the largest known member of the animal kingdom. In our present dimension it appears to be smaller than our forearm, while the human beings viewing it are about the size of the first joint of our pinky. The mass of a blue whale is about 100,000 kilograms, approximately a thousand times greater than that of a large human being.

Living things of greater size than the blue whale exist. If we chose we could spend the better part of a day in walking to California to examine the giant redwood and Sequoia trees. These trees represent the tallest examples of Earth-life, yet would come up only to our waist on level +2. Sequoia trees are also much heavier than whales and are probably the heaviest living organisms, reaching a total weight of several million kilograms, or that of a small ship. Thus, after riding two expanding levels of COSMEL, we have exceeded the upper size boundary of known living things. We will take COSMEL to higher levels later on in the book when we search for other homes for life. For now, we will return to ground level and prepare for our quest to examine the smaller types of Earthlife.

We were able to travel long distances on the higher levels of COSMEL, but on the lower ones we will be confined to a relatively small area. We want to be sure that it contains various things of interest. Luckily, the smaller types of Earthlife are ubiquitous in their distribution, so we don't have to look hard to find them. We choose to explore a wooden kitchen table, measuring perhaps one and a half meters long and one meter wide (Fig. 1). A meal was served there last night, and it was cleaned up somewhat untidily. A small salt shaker and a sugar bowl are on the table. Near one edge there is a drop of chicken soup and some spilled salt and sugar. A mosquito is perched a few centimeters from the drop of soup. This prosaic setting will be the scene of our explorations into what Earthlife is made of. But before we begin, we will move a salt crystal and a sugar crystal until they both touch and are as close to the soup drop as we can manage without their getting wet. We now summon COSMEL.

Before we use it to reduce our size, we want to be sure to emerge at the desired spot. We throw a switch that

Figure 1. The kitchen table to be explored with COSMEL

activates the television screen above the buttons and observe that it displays the kitchen. Using the dials, we can focus the screen on the tabletop and then push the −1 button.

THE WORLDS OF SMALL LIFE

We step out onto a platform comparable in size to a tennis court. In the midst of it the salt shaker and sugar bowl are the size of a fire hydrant and trash basket. The mosquito is a nasty creature which, with legs and antennae extended, is larger than our hand. Peering over the edge of the table, we can see the kitchen floor, which appears to be ten meters down. It is clear that our elevator is functioning well in the reducing mode. But there are many levels that we wish to traverse. We summon COSMEL again, enter, aim the television screen at the area adjacent to the soup drop, and push the −2 button (Fig. 2).

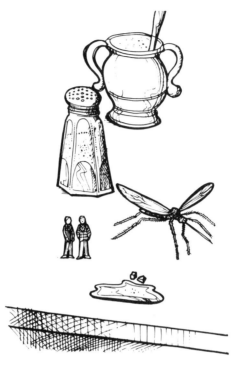

Figure 2. On the table, at the −2 level

The tabletop area is now the size of the playing field in a large stadium. In the distance we can see the shaker and bowl, the size of large statues. The mosquito is formidable and fierce looking, as large as a small automobile. We are glad the insect is only an image, for we would not like to fight it. The salt and sugar crystals look like two blocks from a child's playset. The salt is a white cube, while the sugar has a more angular shape and is translucent. About a finger's length away from these objects is the soup droplet, the size of a large puddle. Several small creatures are easily visible within it. The closest one is somewhat cylindrical, with a tapered end and a group of small tentacles attached to the other end (Plate 2). This creature is called a rotifer, and at level −2, it is about the size of our pinky. Even smaller creatures abound in the soup and tax our eyesight.

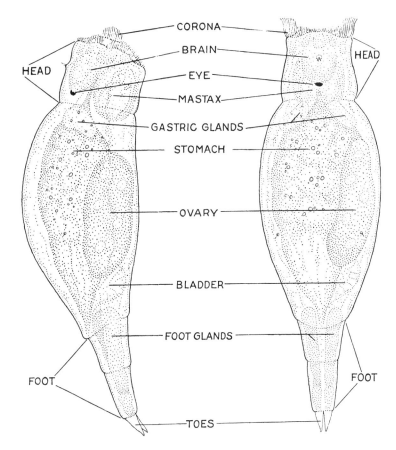

CORONA

BRAIN

HEAD

EYE

MASTAX

HEAD

GASTRIC GLANDS

STOMACH

OVARY

BLADDER

FOOT GLANDS

FOOT

FOOT

TOES

Plate 2. Anatomy of typical rotifer

One, the size of a paper clip, is a paramecium, a large one-celled organism. We recognize hair-like cilia on its surface and a dark nucleus inside it. The single cell of the paramecium is rather large as cells go, although by no means the largest, since an ostrich egg is also a single cell. Most cells are smaller than the paramecium, and we can obtain some idea of how small by straining our eyes and looking at the rotifer, which is also made up of several thousand cells. These cells are similar to those of the paramecium in that each contains a definite boundary and a nucleus. A flake of parsley also floats in the soup, and we can see that this is also made up of cells.

The cell is, in fact, a unit of structure that is characteristic of Earthlife (Plate 3), just as a room is a unit of structure that is characteristic of a building. But whereas the buildings we know contain from one to several thousand rooms, living creatures range in their number of cells from one (as in the paramecium) to many trillions—as in ourselves, for instance. Although most examples of Earthlife are composed of cells, these cells are not its essence, any more than the essence of a city is represented by the rooms of the buildings in it. What is essential in Earthlife is what happens inside the cells, among their components. To study this we will examine cells in detail by using lower levels of COSMEL. We reenter the elevator and focus the television screen at the base of the sugar and salt crystals. We will leap from the −2 to the −4 level. The world will appear as if we were one ten-thousandth of our normal size.

The salt and sugar crystals immediately catch our attention. They appear to be granite blocks about five meters high. If a page of this book were present on the table, it would appear as an enormous slab as high as our waist but over a mile long. We would be about the size of a period in the text, almost as small an object as the human eye can normally see. The rounded edge of the soup drop also towers above us a few meters away. It is now a pond, about the size of a city block. We enter the drop to view the objects within it, beginning with the paramecium, which is now about our size and somewhat transparent. A bewildering array of structures lie within it—membranes, droplets of fluid, tubules, and other objects. When we swim over to look at a cell in the outer wall of the rotifer, we again see many structures within the cell called organelles, some of which correspond to those in the paramecium. Each rotifer cell carries a nucleus, a dark area (in our model) in its own membrane. We can also see within the cell groups of stringy particles re-

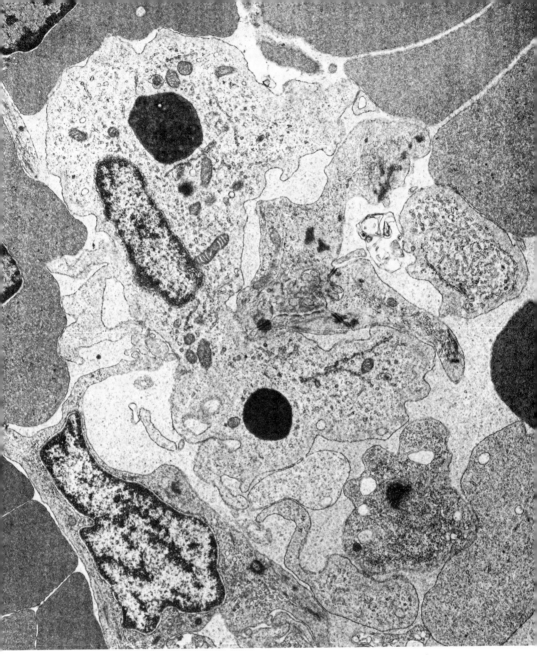

Plate 3. A variety of cell types can be observed, using an electron microscope, in this sample taken from the spleen of a fish. The darker cells which line the periphery of the photograph are red blood cells. Within many of the cells, even darker areas can be seen which represent the cell nuclei. The cells are slightly larger in this view than they would appear to be on the —4 level of COSMEL.

sembling bits of shoelace. These are called mitochondria.

Another object worth examining is the parsley flake. Its leading edge is as thick as we are tall. The green color is associated with cells in the interior and localized in button-shaped particles called chloroplasts within the cell. These particles contain a green pigment called chlorophyll. We will later see that chlorophyll plays a key role in the process by which some living things gather energy. In addition to the chloroplasts, the cells in the parsley also contain many of the other structures that we have already encountered in the paramecium and rotifer.

It is clear from what we have seen so far that even at a level that is ten thousand times smaller than usual, Earth-life is still very complex, and we have not yet found its essence. We must proceed deeper into a microscopic world to locate the smallest forms of Earthlife, and to explore the fundamental aspects of how both tiny and large living things function. We can see some still smaller living things in the same drop of soup—in particular, a cluster of objects, each the size of a finger joint. (These minute shapes, called bacteria, were first observed in the seventeenth century.) To inspect them properly, we will use COSMEL to adjust our size to that of a bacterium. We summon our elevator to us, within the soup drop. With the television screen centered on the group of bacteria, we push the button marked -6 and step out again.

The bacteria are now each about the size of our own body. In reality, they measure about a millionth of a meter long, or thirty million times less than the length of a Sequoia. The weight of a bacterium is roughly 10^{-12} grams, which means that a whale is in weight to a bacterium as the whole Earth is to the whale. These bacteria belong to a species called *Escherichia coli,* which is normally found in our gut but occurs also in food. At this level we can see that a half-dozen long, whip-like filaments (called flagella) project from each bacterium. They are no thicker than our thumbs, but extend for about seven meters and are obviously useful in locomotion. We can be grateful again that we are observing a model. If the bacteria were in motion at normal speed, they would appear to be traveling at about sixty-five kilometers per hour. Also if we were actually their size, a problem would arise. Our ability to view small objects by visible light is itself limited, and we could not see the details of objects smaller than our forearms on this level. In the laboratory, microbioligists have circumvented this limitation by using a device which employs a different type of illumination that can image smaller objects—the electron microscope. In our model we have avoided the problem by adjusting the image of the bacterium to human dimensions.

We have arranged in our model for the outer coverings of each bacterium to be transparent. Therefore, we can peer through the rough outer cell wall (about a finger's length in thickness), as well as through the flexible cell membrane, which is of equal thickness, that lies directly beneath it, and see the contents of the bacterial cell. We notice that many of the structures observed in larger cells on the −4 level are absent here. There is no nucleus wrapped in its own membrane, no mitochondria, and no chloroplasts. Any of these would fill the entire volume of a bacterial cell. These structures are obviously not essential to life. But even the bacterial cell is by no means a simple entity. Each flagellum, for example, is anchored to the cell wall by a complex device composed of at least a half-dozen rods, discs, and connecting pieces. Thousands of roughly spherical particles, each about the size of a quarter in diameter, can be seen in the cell fluid. They are called ribosomes and occur within the cells of every known creature on Earth. We will soon see that ribosomes play an important role in the fundamental life process.

One other structure that can be seen in the bacterium is particularly important: It resembles a shoelace wrapped and rewrapped in loops around a core to form a fairly compact ball. It is attached to the inside of the cell membrane. As we shall see, the chemical, DNA, that composes the shoelace is crucial to the functioning of all forms of Earthlife with which we are familiar. The structure that contains the chemical is called the bacterial chromosome. If we inspected the cells of the rotifer, paramecium, parsley, or mosquito on the −6 level, we would again detect DNA, but in the higher organisms the DNA would be separate from the cellular fluid. It would be found primarily in the nucleus, but also to some extent in the mitochondria and chloroplasts. Also, there is up to a thousand times more DNA in the cells of some higher organisms than in those of a bacterium.

Let us turn our eyes away from the *Escherichia coli* bacterial colony and take note of other objects in the soup around us. We notice immediately that *E. coli* is not the smallest type of bacteria present; the one that we observed was about our own size. Other bacteria nearby, however, range down to a size approximating a large beach ball. Each of these creatures has a cell covering, and a fluid substance within which contains ribosomes, a chromosome containing DNA, and other structures. As far as we can see, there are no cells that are much smaller, but other structures are present in the soup that are complex and may be of biological origin. One particularly striking one appears as we have indicated in Fig. 3.

In this specimen a head, hexagonal in cross section, sits

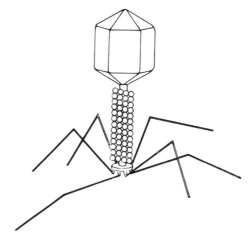

Figure 3. A T2 virus

upon a shaft of equal length. Six jointed, spindly legs project from the far end of the shaft. Suitable small connecting pieces join the shaft to the head and legs. The head is about the size of a fist, and with legs extended the entire apparatus is about the length of a forearm. This structure has the appearance of a toy model of a Moon lander vehicle, and on the whole, looks more like a mechanical contrivance than a living thing. As we take a stroll through the colony of bacteria, however, we notice a number of these particles in the water and, finally, one that is attached by its legs to a bacterium. Another such particle sits on the surface of a second bacterium, but in this case the legs have contracted so that the end of the shaft is in contact with the bacterial cell wall. Finally, we observe a slaughtered bacterium with its cell wall split open. A number of the "Moon lander" particles are emerging from the inside of the split bacterium.

These objects are, in fact, part of the realm of life. The ones we have described are viruses; in this particular case, it is called the T2 virus. It is a bacteriophage, a virus that is parasitic on bacteria.

While the shape of the T2 virus brought it to our attention, it is not the only virus present. On further inspection we see a number of others that are smaller than T2. They are present as rods, spheres, and other shapes, and the smallest is no bigger than a quarter in cross section.

These small viruses represent the size limit of known things parasitic on Earthlife. The one with the smallest mass is probably a viroid, a rod-like object which on the -6 level would appear to be a few centimeters long. The viroid is not present in the soup drop, as it is known to be a parasite only on certain plant cells, although some animal diseases, such as scrapie which infects sheep, may be caused by substances related to viroids. These viroids are in size to a bacterium about as a flea is to a small dog. Viruses are not independent living things in the same sense as bacteria, since they can perform some of their important functions, such as reproduction, only inside a cell of another living thing. But such dependence of one living organism on another is not uncommon. For example, cows cannot digest their solid food, grass, without the aid of some bacteria that live in their digestive tracts. Furthermore, the chemical substances that play essential roles in the life of bacteria and higher organisms—nucleic acids and proteins—play similar roles in the activities of viruses. It therefore seems unreasonable to deny that viruses are living just because they need help to do so.

In order to reduce our size to less than that of the smaller viruses, we would have to descend to the −8 level. Thus, in finding the largest and smallest forms of Earthlife, we have had to travel through ten levels of COSMEL. This is a much wider range than is needed to explore the world of buildings, in which the Empire State Building would be only two levels above the simplest shack. Our everyday experience takes place at ground level, that of our familiar environment. We humans are among the largest creatures of Earthlife, with many more populated levels below us than above. If a bacterium had the capacity and inclination to use COSMEL, it would have a very different point of view. It might choose to renumber the levels so that the one we call −6 would be termed ground level, and the world of our everyday experience would be labeled the +6 level. We shall see later on that a quite different perspective exists on the cosmic scale as well, in which human beings are closer in size to the smallest objects we know in the Universe than we are to the largest ones.

Our examination of the creatures of Earthlife, large and small, has indicated that they are complex, somewhat like a Chinese puzzle box. But we have not yet found out what fundamental structures and principles allow them to live. In order to study this, we will descend to still lower levels and look at the objects known as molecules and atoms of which both living and nonliving matter is made. Here we will finally see how the activities of Earthlife are elaborations of the properties of two special types of molecule—

nucleic acids and proteins. But first we will examine some atoms and molecules in nonliving matter, the salt and sugar crystals. While we are still on the −6 level, we will make the journey over to them to take a look. We will therefore step out of the soup droplet (now the size of a ten-kilometer lake) which has been the site of our activity on this level so far. The two crystals appear in the distance as twin cliffs, five hundred meters high and about the same distance away. A moderate hike across the plain brings us to the point where their bases are in contact. We will reenter COSMEL at this point for the trip to the −8 level.

WHAT LIFE IS MADE OF

We must pause for some explanation before we step out at this level. At each of the higher levels that we visited, we saw new perspectives, shapes, and forms, but with some practice we could interpret the new world we saw in terms of our familiar one. A paramecium or bacterium was perceived as a beast of conventional size but strange form. The salt crystal was a white block that varied in size from level to level, but it was still a white block. In constructing a model of the −8 level, we must introduce a greater number of adaptations and distortions in order to keep the environment comprehensible to us. At this level the apparent solid facade of matter separates into component atoms in a manner similar to that in which a newspaper photograph, viewed with a magnifying glass, resolves into a series of white dots on a black background and black dots on a white background. Atoms consist mostly of empty space and do not have definite borders. Later we will look at them in this greater detail. Nevertheless, in our model of the −8 level, we have represented atoms by solid, roughly spherical objects that vary in size from a marble to a Ping-Pong ball. They look similar to the plastic models of atoms used by chemists in the laboratory. Just as the chemist does, we have used different colors to distinguish atoms of different types.

Real atoms do not have any permanent colors, although under some conditions they emit light of various colors. Furthermore, we have provided the model with a number of switches that control the lighting effects. We can thus make whole sets of atoms transparent as we choose, so that others can be observed. With this preamble we are ready to step out of COSMEL and into the world of level −8.

At first glance we see a number of objects suspended in space before our eyes. They are about as big as a marble and on the average, they are about half a meter apart, but their distribution is random, not regular. Some of them can be seen in the process of impending, or actual, collision. These objects are the particles of which our air is composed. We have made time stand still. These objects, if in normal motion, would move at a rate too fast for us to follow.

In our frozen-time model, we can inspect the air particles at our leisure. We notice first that almost all of the particles consist of two atoms stuck together. Of the rest, about 1 percent contain a single atom, while another 1 percent have three atoms. Furthermore, the particles (we will use the word "molecule" from now on) containing two atoms are of two types. The lesser component, oxygen (about 20 percent of the air), has two atoms which we have colored red. This color was selected to suggest its vital role in sustaining fires. As we shall see, it has a similar function in the processes of many living organisms on Earth. The major component, nitrogen (almost 80 percent of the air), has two atoms stuck together, which we have colored blue.

The type of atom that occurs in air as a single unit is called argon. It is not important to terrestrial life, as it is chemically inactive at Earth temperatures. The molecule of three atoms, present at about 1 percent in our model, has atoms of two different kinds. A central red oxygen atom is flanked by two white ones, representing hydrogen. This substance is water, represented chemically by the formula H_2O. Unlike the other substances we have mentioned, the number of water molecules present can vary considerably from place to place and time to time. On ground level we call this effect a change in humidity. We have represented relatively dry air in the vicinity of the salt crystals (which can absorb water). More water molecules can be expected in the air close to the drop of soup, since they have evaporated from it.

The four molecules mentioned represent nearly all of those present in our air. To find other types, we will patiently have to examine several thousand molecules. Yet we will make this search, for there is another molecule in the air that is important to life. After a few minutes we find a specimen in which two red oxygen atoms flank a central black one. The atom colored black is one that is crucial to Earthlife, carbon. Its color suggests one of the forms in which we encounter carbon in our everyday world—charcoal. This molecule is carbon dioxide. While relatively rare in the air, it occurs in higher concentration in our breaths when we exhale. Carbon dioxide is also the gas that produces the fizz in "carbonated" soft drinks and champagne.

Now that we have examined the components of air, we do not wish those atoms to distract us from further observations; therefore, we will alter the lighting scheme so that the air molecules become invisible. We immediately see two gigantic walls that tower above us to heights beyond our perception. These walls are the faces of the salt and sugar crystals, and on this level they would reach a height of five Mount Everests.

Turning our attention first to the salt, we find that this wall is composed of two kinds of atoms in a perfectly regular, alternating, two-dimensional array (Fig. 4). The face is not exactly flat because the two different atoms are not the same size. The atoms are sufficiently transparent for us to perceive that this arrangement also extends back into the wall, in the third dimension. We have chosen the color gray for the smaller atom, sodium, to accord with its metallic character. The larger one, chlorine, has been colored green because when it occurs as a pure gas on ground level, we

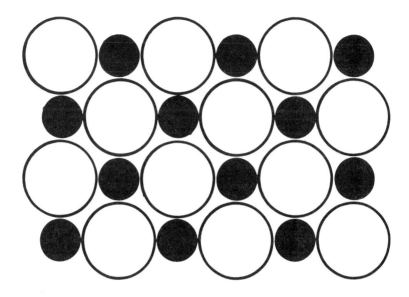

Figure 4. A portion of the face of the salt crystal. The small dark circles represent sodium atoms, and the larger, light ones are chlorines.

perceive it as green. The atoms that comprise this gigantic salt crystal are fixed in place and are in contact with each other on all sides. This arrangement is characteristic of a solid, but quite different from the random arrangement of molecules present in the air, which is a gas.

We walk a few paces over to examine the face of the other wall, which is made of common table sugar. We can detect the presence of red, white, and black atoms, representing oxygen, hydrogen, and carbon, respectively. Their arrangement is regular but much more complex than the simple alternating scheme present in the salt. The sugar wall is a huge stack of identical irregular units, each of which is an individual sugar molecule, which is roughly the size of our hand. A count reveals that forty-five atoms are present in each unit. The sugar molecule is thus larger than the air molecule or the simple repeating unit of salt. On closer inspection we see that the carbon atoms are located most centrally in the sugar molecules, with the hydrogen atoms on the outside. There are two hydrogens to each oxygen, the same ratio that is present in water. The term carbohydrate has been applied to a class of substances that includes table sugar, because a count of atoms suggests it is composed of carbon and water. This is of course not literally true. No water molecules are present. Instead the forty-five atoms in the molecule are interconnected in a rather involved way (Plate 4).

Plate 4.   Model of a molecule of table sugar (sucrose). The white atoms represent hydrogens, the black ones in the interior represent carbons, and the ones of intermediate color, oxygen. The slots on atoms of oxygen do not represent structures present at the atomic level, but are used as an aid in constructing models. The size of the molecule on the page is slightly larger than it would appear on the −8 level of COSMEL.

A number of sugar molecules have become detached from the wall and lie on the ground at our feet. (At the same time we notice of course that the table surface itself is a solid composed of atoms. For the time being, we adjust our lights so that this surface provides a uniform background; we will inspect it in detail in a moment.) We are struck by a particular property of the sugar molecules. They are not only the size of our hand on this level, but they share another feature in common with a hand—they are not symmetrical. In examining our hands, we note that one is approximately a mirror image of the other, but we cannot make it occupy the space occupied by the other one. You can test this by trying to place your right hand into a baseball glove meant to fit on the left one. Each sugar molecule, as a whole, is also asymmetrical. Therefore, we might expect to find left- and right-handed sugar molecules which are mirror images of each other but otherwise identical in the way in which atoms are connected together. But as we scramble about inspecting one sugar molecule after another, we eventually make a surprising discovery. Only one form of sugar molecule is present. We will refer to it as right-handed. The mirror-image molecule could be made in the laboratory, but it is not found in living things on Earth. We will see that a similar situation occurs for other molecules important to life.

Let us next pay attention to the "floor," the wooden surface of the table. An array of hydrogen, oxygen, and carbon atoms, white, red and black, stretches into the distance. The atoms are not divided into discrete molecules, however, but appear interconnected in a continuous carpet without interruption. The arrangement is not chaotic; we can see that repeating subunits are present and arranged into very long chains. Chemists call such an arrangement a polymer. On examining one subunit (called a monomer) we note that it is a carbohydrate with a fairly close resemblance to sugar.

Our principal aim in making the trip to the −8 level has been to examine living things and not just their products such as sugar and wood, so we will return to the drop of soup. To do so on this level would involve a hike of fifty kilometers, but we can speed up the process by using COSMEL as a shuttle. We will return to the −6 level, retrace our steps (now only 0.5 kilometer) to the soup drop, reenter the drop, locate the same bacterial colony, and then return to the −8 level. We do this and on opening the elevator door, find that a "sea" of red and white water molecules completely fills the doorway. We notice that molecules in the liquid state, as in the solid, are in close contact but are not nearly so orderly. We can allow time to elapse very slowly, as in a slow-motion television replay, and see the molecules rearranging themselves and pushing past one another much like a crowd in Times Square on New Year's Eve. Like people in a crowd, a few molecules in the liquid may stick together in an ordered way, but they do not remain together with other nearby, but unrelated, molecules. As we do not wish to examine the water molecules further, we adjust our lights to render them transparent (as if we were in a lake of water at ground level) and step out to inspect the contents of the soup.

The scene resembles the one we saw earlier in the air. A number of atoms and molecules are suspended before us. Some are familiar; we recognize red oxygen molecules and black and red carbon dioxides. Gray sodiums and green chlorines are there (the soup is salty), but they are not arranged in a regular pattern as in a salt crystal. Many larger molecules, composed of up to perhaps one hundred atoms, are present. We can detect our now-familiar table sugar, plus a number of smaller carbohydrates. We also note a large variety of molecules which we have not seen before that display the blue of nitrogen, in addition to the red, white, and black of oxygen, hydrogen, and carbon. These molecules are important and are found in greater abundance inside of living things.

A bacterium looms directly in front of us, the size of an ocean liner, with a T2 virus attached to it. At level −8, the virus is no longer a toy Moon lander but is larger than the authentic item. It extends about twenty meters even in its contracted form. It invites our inspection, but we will pass it by for now. The bacterial cell wall in front of us presents a complex surface about a meter thick, which is made of molecules containing carbon, hydrogen, oxygen, nitrogen, and other atoms. We pass through it quickly and also pass through the cell membrane, another complex thick structure containing several layers, in order to devote our attention to the contents of the cell (Table 2-1).

The cellular fluid contains water as well as a rich array of other molecules of various sizes and shapes. We have encountered many of them in the soup outside, but not in this great density. Obviously, one function of the cellular covering is to prevent this rich, thick broth of molecules from escaping into the thinner soup outside. The thick molecular broth is an important part of the life of the bacterium, but it is not yet the essence of it. It is as if we had found the instruments of an orchestra but not yet located the musicians. There are also some larger shapes and forms present in the cell. We will focus our attention on the larger ones (macromolecules), of which there are many fewer.

The moment we do this we become aware of a number of roughly spherical irregular shapes about us that vary in size from a basketball to a small chest. Each contains thousands of atoms. Many different types are present, but we occasionally see several macromolecules that are identical. All the ones that surround us have, however, a common pattern to their construction: It is as if a long chain had been wrapped and folded to make an irregular three-dimensional object (Fig. 5). Our use of the word chain is also appropriate because we have here another example of a polymer. Simple units have been connected repeatedly to make a long array. The links are not identical, however. Each contains a part, or marker (one of twenty), that distinguishes it from its fellow. A diagram can clarify this construction (Fig. 6).

The chain is shown, and the first twenty letters of the alphabet indicate the different markers. The whole situation resembles a railroad train made up of twenty different types of car. Not every chain contains all twenty units, but if we survey every chain in the cell, we would eventually find a

Table 2-1

THE CONTENTS OF A BACTERIAL CELL *

Substance	Number in Cell	Fraction of Weight of Cell
Small molecules		
Water	40 billion	.70
Salts	250 million	.01
Carbohydrates (small and large)	200 million	.03
Amino acids (not in proteins)	30 million	4×10^{-3}
Nucleotides (not in nucleic acids)	12 million	4×10^{-3}
Fats	25 million	.02
Other small molecules	15 million	2×10^{-3}
Large molecules		
Proteins	1 million	.15
DNA	4	.01
RNA	460 thousand	.06

* The contents of a bacterial cell while it is in rapid division.

total of twenty different kinds. The type of molecule that we have described is called a protein. The individual units used to construct a protein are called amino acids. We have seen amino acids among the small molecules that were plentiful both in the soup and the bacterial cell fluid. Amino-acid molecules, like table-sugar molecules, occur in living things in only one of the two possible mirror-image forms (Fig. 7).

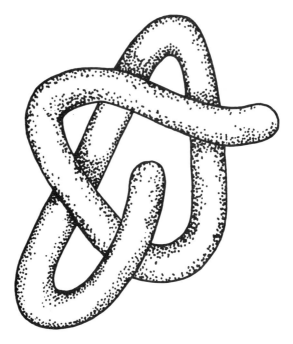

Figure 5. The three-dimensional outline of a protein chain

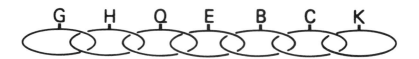

Figure 6. A symbolic representation of a portion of a protein chain

As we look about the cell, it is obvious that proteins play many roles in the bacterium. They are present in the cell wall and membrane. One of the bacterial flagella is anchored in the wall close to us. We observe that the various parts used in its construction are also proteins. We understand the role played by these structural units. However, the function of the globular balls that hang in front of us is less obvious, so we will examine them in more detail.

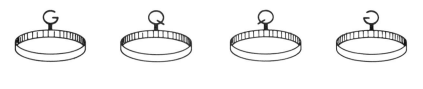

Present   Not present

Figure 7. A symbolic representation of amino acids, units of the protein chain. The units (left) and their mirror images (right) cannot be superimposed upon each other. Only one of each pair of mirror image forms occurs in proteins.

If we compare two protein molecules that are identical in size and shape, we find that the order of the links along the chain is the same in both cases. Different proteins have their amino acids linked in different orders, or contain a different subset of the twenty units. The atoms present in proteins are for the most part those we have already met: carbon, hydrogen, nitrogen, and oxygen. An additional type of atom present, colored bright yellow, is called sulfur; its color is the one sulfur has in our ground-level world. We note that sulfur atoms are used in pairs within a protein to lash together parts of the chain that are remote from each other in their linking order. This serves to hold the molecule fast in a particular three-dimensional shape that is involved in one of the most important functions of protein—to help chemical reactions take place. By allowing time to run forward slowly, we can see how a protein does this. We observe a small molecule approaching the surface of the protein. It enters a particular cleft on the surface that just fits its shape, and emerges, splits into two smaller fragments. The opposite process, two small molecules combining into one large one on the protein surface, also can occur.

When we examine other protein molecules, we find that similar things are taking place. Each kind of protein seems to help carry out (catalyze) a particular transformation of one or more small molecules. It need not always break a bigger molecule into smaller parts. It may catalyze a change in shape of a single molecule or an exchange of parts between molecules. The name given to this class of proteins as a whole is enzymes. Enzymes are not the only possible catalysts. Many other substances, such as small particles of certain metals, can also act catalytically. But proteins are among the most effective known catalysts in terms of how much they can speed up the rate of a chemical reaction. Also they tend to be very specific in their actions, affecting one or a small number of reactions while having no influence on others. They are able to do this because of their complex three-dimensional shape, which allows some molecules to enter while barring others.

As we wander through the interior of the bacterium, it becomes clear that a vast array of hundreds or perhaps thousands of enzymes are performing a large number of processes that are vital to the life of the organism. Amino acids and parts for the cell wall and membrane are being constructed, carbohydrates disassembled, the necessities of life prepared, repaired, and stored. The various steps are interrelated in a manner resembling the processes in a vast factory. The finished product from one reaction is the raw material needed for another one, either close at hand or at a distance in the cell. The action of the enzymes is vital in keeping materials moving through this series of interlocked cycles. We can get an idea of the complexity of the cycles by looking at a chart which outlines a portion of known cellular metabolism (Plate 5).

But where are the enzymes themselves constructed? Our path eventually takes us to a group of structures that we first observed at the —6 level, the ribosomes. At our present level each ribosome is a large, vaguely heart-shaped object, about as wide in diameter as we are tall. From a cleft on the nearest ribosome, a partly constructed protein chain emerges, and starts to fold into a three-dimensional shape. The ribosome, then, is the factory for making proteins and a vital component of the living process.

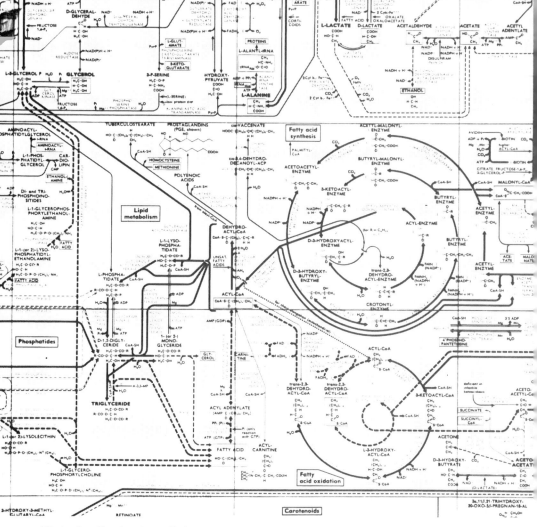

Plate 5. Cycles in cellular metabolism. The section displayed represents about 12% of the area of a large chart, which represents the known biochemical pathways in a typical mammalian cell.

What is the ribosome itself made of? Much of it is several different types of protein stuck together. But the majority of it appears to be made of a different type of macromolecule that twists in and out among the various proteins. This new type of molecule is again a polymer, consisting of repeating subunits. An atom which we have not encountered before is present. This is phosphorus, which we will color orange. Every orange phosphorus atom is surrounded by a cluster of red oxygen atoms to make a group called phosphate. In the polymer, phosphates alternate in a chain with a subunit that is a smaller relative of table sugar. We will simply call it "sugar" and trust the reader to remember that it differs from our familiar table sugar. Apart from phosphate and sugar, our new molecule also contains a marker that distinguishes each subunit. Protein chains contained up to twenty different markers, while this molecule (called a nucleic acid) has only four, which are different from the ones in protein. A simple symbolic representation of a portion of a nucleic acid chain is shown in Fig. 8.

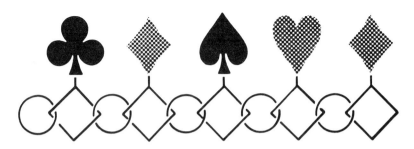

Figure 8. A symbolic representation of a portion of a nucleic acid chain

In this figure, circles represent phosphates, squares represent the sugars, and the card-suit symbols represent the four markers (which chemists call bases). The separate subunits of the nucleic acid chain were among the small molecules that we saw floating in the bacterial fluid. Such a subunit is called a nucleotide. (See Fig. 9.) When we follow the order of the bases along a nucleic acid chain, it appears to be random; we can detect no regular repeating units. However, when we examine many different nucleic acid chains in a single bacterium, we find that there is a lot of repetition. The same long chains are present in every ribosome. It is like being dealt the same hand at bridge for many games in a row. Also, if we were to examine different bacteria, we would again find similar, although perhaps not identical, nucleic acid chains in them, analogous to many

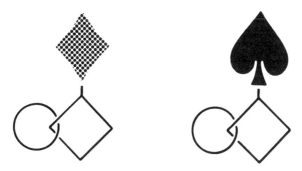

Figure 9. A symbolic representation of nucleotides

players at different tables getting the same hand. Clearly, something interesting is involved here, and its study will take us closer to the essence of Earthlife.

Now that we can recognize the structure of a nucleic acid, it is evident that other smaller ones are present in the fluid near the ribosome. One very common type has about eighty subunits folded into an L-shaped molecule (the ribosome contained thousands of nucleic acid subunits). About half of these L-shaped nucleic acids contain an amino acid attached to their ends, while the remainder have none. These smaller nucleic acids are transporting amino acids to the ribosome as raw materials for the construction of proteins. As we have said, each ribosome is a factory in which a protein molecule is being assembled from amino acids. But what determines which protein is being created by a particular ribosome?

To learn more we will step back and examine the entire cluster of ribosomes. They are connected in a series by a long nucleic acid molecule, the thickness of a rope, which passes through each ribosome. (Plate 6.) The cluster resembles a gigantic assembly of beads on a string. Thus, nucleic acid molecules serve both as part of the structure of the ribosome (the bead) and as the string which connects the beads. Each ribosome along the string is manufacturing the same protein, while ribosomes along different strings are making other proteins. We can locate the free end of this string and walk along it. The farther we progress from the end, the more

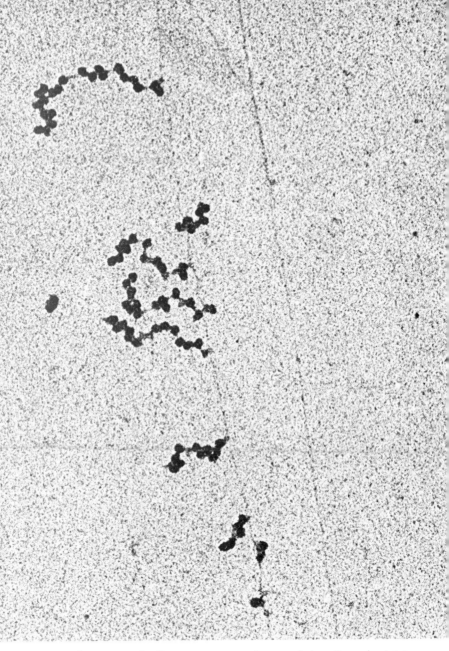

Plate 6. Clusters of ribosomes, as observed by the electron microscope. The thin, threadlike molecule to which the various bead-on-a-string arrays are attached is called DNA. The magnification used renders them slightly larger than they would appear on the −5 level of COSMEL. At the −8 level, each of the individual ribosomes would be comparable in size to a human being.

complete the protein that is being manufactured in the ribosomes we pass along the way. If we were to take the time to mark down the arrangement of the four bases along the nucleic acid string and compare it with the sequence of the twenty amino acids in the protein being made, we could deduce that there is a relation between the two sets of symbols.

An analogue of the process is shown in Fig. 10. The code used is not very elaborate, as it simply converts lower-case letters in the nucleic acid to capital letters in the protein. There need not be such a close correspondence, however, and the message could be translated into a totally different set of symbols as well. This is what occurs in Earthlife. Biochemists have deduced the relation between the nucleic acid and protein languages (the "genetic code") by comparing a variety of nucleic acid code sequences with the amino acid series in the proteins that are produced when the sequences are "read" by ribosomes. They have found that a particular group of three bases (a triplet) along the nucleic acid string specifies a particular amino acid along the protein chain. Other triplets indicate where to start and stop making the protein. For this reason the nucleic acid string which held the ribosome beads is called a "messenger." But where, then, does the message originate?

We can locate the source of the information simply by walking to the other end of the messenger string. We find that it leads to the molecule called DNA to which one end of the messenger is attached. The area of their attachment somewhat resembles a railroad switching junction. The messenger chain runs alongside a DNA chain for a time and then terminates. Its end is marked by a protein. The DNA chain runs alone past this point for a distance and then is joined by another DNA chain. Thereafter, they continue together. The entire assembly has groups of proteins attached in various places.

We can begin our examination at the simplest point— where the DNA chain runs alone. We see that DNA is a nucleic acid, containing sugar, phosphate, and base units.

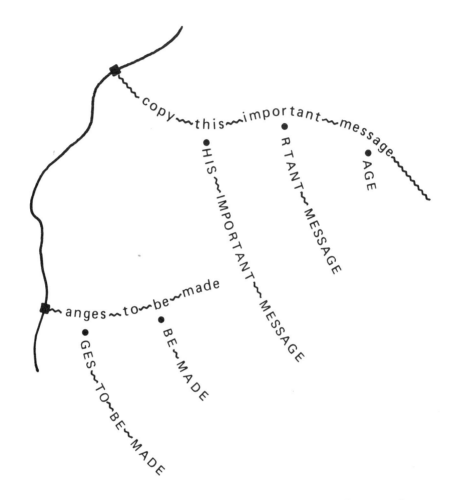

Figure 10. A symbolic representation of the manufacture of proteins. The dark line represents DNA, and the dark square is a protein involved in the manufacture of a messenger molecule (wiggly line containing small-case letters. In our symbolism, each small letter represents three nucleotide units). The dark circles (ribosomes) move along the messenger toward DNA. As they do so, the message is copied into a protein chain (the string of capital letters, each representing an amino acid). The second message, "there are changes to be made," is still in the process of being manufactured, but copying already has begun.

Two minor chemical differences distinguish it from those nucleic acids present in the ribosome, those that act as the messenger and those that serve as the L-shaped transport for amino acids (these types are called RNA). An alteration has been made in the sugar as well as in one of the four bases in DNA, but otherwise RNA and a single strand of DNA are very similar molecules. The main difference is in the number of subunits, with the bacterial DNA containing many times more than the RNA. In size they compare somewhat as this book would to a single page, with the DNA containing a few million nucleotides, the RNA a few thousand.

Next we examine the place where two DNA chains run together. The chains are not merely parallel, as our railroad-track analogy might imply, but twisted intimately around one another to form a structure called a double helix. At the -6 level we observed the helix to be the thickness of a shoelace; now at the -8 level it is about as thick as our upper thigh. The bases of each DNA chain are in the most central position of the helix, with those of one chain in actual contact with those of the other. The sequence of bases on one chain is related to the sequence on the other. This code, fortunately for our egos, is simple enough for us to crack on the spot. The bases occur in pairs. We can illustrate this with the card-suit symbols we used before; each symbol faces the other of the same color. A club always appears across from a spade, a diamond from a heart (Fig. 11). When you know what is along one chain, you can immediately write down what is on the other.

Finally, we return to the junction where the messenger RNA chain meets the DNA chain. The manner in which the DNA and RNA chains run together is very similar to the way in which the DNA chains united; a double helix is formed. The protein that is present at the end of the messenger RNA chain contains in a cleft a small molecule. We recognize this as a nucleotide, but it contains two extra phosphate groups (called a nucleoside triphosphate) as illustrated (Fig. 12). When we allow time to elapse, we can see

Figure 11 A symbolic representation of base pairing between two nucleic acid chains

Figure 12 A symbolic representation of a nucleoside triphosphate

the messenger actually being manufactured at this point, and the information defined by the sequence of bases along it is being copied from the DNA chain. The nucleoside triphosphate on the enzyme is a subunit waiting to be built into the messenger. In this process the extra phosphates are removed, just as we remove the cap before consuming a bottled soft drink. So the proteins that make up much of the cell and do its chemical business are created in the ribosomes, according to instructions that originate in the cell DNA, and are carried to the ribosomes by messenger RNA. Where then does the information in the DNA for doing this come from? We now let time move forward and watch the DNA for a while. Eventually the bacterium prepares for its ultimate fate—cell division. There is a flurry of activity, full of complex maneuvers in the vicinity of the DNA. Again we find a situation that resembles a railroad switching yard. The two "tracks" of DNA separate, and alongside each, a new "track" is constructed. The process used is similar to the one described above in which an RNA chain was constructed alongside of a DNA chain. The subunits used are again nucleoside triphosphates. The sequence of bases in the existing chain controls the sequence in the chain that is being built, according to the pairing scheme illustrated with symbols in Figure 11, above. In the end two complete DNA double helices are made, each containing one component of the old one. This process is known as replication. The bacterium then divides, and each new bacterium contains one of the double helices of DNA. The answer to our last question is then: The information in DNA is copied from previously existing DNA. It is this information that is the main legacy of a cell to its descendants. Of course, this raises an obvious question: Where did the first DNA come from? As this is a question about the history of Earthlife, we will postpone discussing it until Chapters 4 and 5.

We are still standing inside a bacterium in the vicinity of its DNA at the −8 COSMEL level. On this level we will now make one additional detour to inspect the Moon-lander virus sticking to the bacterial surface.

We can see that the legs, shaft, head, and connecting parts of this virus are all proteins. Only after we adjust the lights to render the proteins transparent can we detect an additional chemical substance. A length of double-helical DNA lies folded within the head of the virus. By watching a while, we can reconstruct the life cycle of the virus. The DNA is injected down the shaft, through the cell wall, and into the bacterium. It alone enters the cell. Once inside, it borrows the ribosome, enzymes, and small molecules of the bacterium. The viral DNA is copied, and the proteins necessary to assemble additional Moon-lander virus particles are constructed. An enzyme is made that destroys the bacterial cell wall. The bacterial cell ruptures, and the newly assembled T2 viruses escape to seek new victims. Because the viral DNA does not contain instructions for making most of the cell components, it is relatively small, containing a few hundred thousand subunits. The viroids that we mentioned above are even tinier, being made of a small piece of nucleic acid containing just a few hundred subunits. The viroid nucleic acid chain reverses direction at its midpoint, and bends back to form a double helix with itself. The ends of the chain are connected; therefore the entire piece of nucleic acid is circular. This small, circular nucleic-acid molecule is in itself a functional unit of Earthlife, sufficient to cause a disease that blights large organisms such as plants and trees. It is the least complicated thing that we know which carries its own genetic information and can be considered as a separate functional unit.

We might continue our journey by examining other creatures in the drop of soup: the rotifer, the parsley, or the paramecium. In each case we would discover many elaborate and ingenious chemical and physical mechanisms. In the end we would find that what happens in the cell is controlled by enzymes, and that the enzymes are manufactured in ribosomes, using instructions originating in DNA. While the specific enzymes and the number and arrangement of bases along the nucleic acid chains vary from bacteria to rotifer to whale, their general pattern is the same, and it is the same in the cells of our own bodies. We have found the essence of Earthlife in this mutual influence of protein and nucleic acid. One acts as catalyst and structural material, the other as a blueprint. Many questions suggest themselves at this point. How did the complex mutual interplay of protein and nucleic acid arise? Could other analogous combinations act as the essence of different types of life on Earth or elsewhere? We will, after two short detours, examine these questions in later chapters.

WHAT EVERYTHING IS MADE OF

The smallest building blocks of Earthlife observed on our tour have been atoms and molecules. They are the structural elements of Earthlife but not, as was once thought, the simplest components of matter. Atoms are themselves made of smaller units. A knowledge of the detailed subatomic structure of matter is not crucial to an understanding of how Earthlife functions, but it may well help us grasp the basis of other kinds of life. Therefore, since we have already traveled far down toward the smallest constituents of matter, let us continue our trip further to learn what atoms are made of.

We select a single water molecule from our soup droplet, center it on our COSMEL screen, and push the −12 button. When we step out of the elevator, we stand before the oxygen atom, which appears as a huge diffuse cloud, a hundred meters across. There is a problem in displaying an atom at this level. Atoms are mostly empty space, and the cloud represents the average location of several bits of negative electricity—the electrons. These electrons normally whiz about at great speed like so many hummingbirds, but this motion is too rapid to follow and indeed, according to present-day physics, it is intrinsically impossible to do so. We have therefore used the cloud to represent the region over which the electrons move.

The two hydrogen atoms of the water molecule can be seen as additional clouds, slightly smaller than the oxygen. They blend into the oxygen cloud so that the water molecule forms a larger bent shape, perhaps two hundred meters from end to end. This larger cloud is inhabited by ten electrons in all. Some of the electrons move only within the center part of the cloud (the oxygen atom), others move from the oxygen area to the hydrogen area and comprise the chemical bond that unites the atoms. If we broke apart a water molecule and viewed the oxygen atom in isolation, we would find that it contained eight electrons, whereas the hydrogen atom has only one electron. It is this difference in electron number that gives the two elements their different chemical properties.

These electrons occupy the entire area of the atom, but comprise less than one-tenth of 1 percent of its mass. Where then is the rest of the mass of the atom and what holds the electrons, which electrically repel one another, together? To find the answer, we must enter the cloud and walk fifty meters to its center. If we search carefully, we will find a tiny spot, measuring about one millimeter, on level −12: the atomic nucleus. We have made it very bright, to aid us in our search and to represent the fact that it is a positive bit of electricity. Although the word "nucleus" is used here, this object is not directly related to the nucleus of a cell or to nucleic acids. The nucleus of the atom holds the electrons together because opposite electric charges attract, and it is thousands of times heavier than the electron.

In order to learn more about the structure of this nucleus, we reenter COSMEL and descend to level −15. The oxygen nucleus now appears as a fuzzy ball with a diameter slightly greater than our height. On this level the nucleus of one of the neighboring hydrogen atoms of our water molecule would be about one hundred kilometers away, somewhere near New Haven if we were in New York City! The oxygen nucleus has a structure. It is made up of two kinds of objects, which we represent as smaller fuzzy balls in our model. One is electrically charged and known as a proton, while the other is neutral (uncharged) and called a neutron. The most common type of oxygen nucleus has eight of each, which are held together in a mutual tight embrace by strong, nonelectrical forces. If we were to examine the nuclei of other atoms, we would find that they differ from each other in that they contain various numbers of protons and neutrons. The hydrogen nucleus, for example, consists of a single proton and no neutrons. Carbon (in its most common form) has six protons and six neutrons.

So all nuclei are made of protons and neutrons, which together with electrons are the constituents of ordinary matter. The electrons, being very light, can go from atom to atom in chemical reactions. In some situations, such as in metals or the interior of stars, they can move about freely quite independently of atomic nuclei. Electrons are ordinarily stable and do not change into anything else. As far as we know at present, they do not have any complex internal structure.

The nucleus of an atom will often be surrounded by just enough electrons to balance its positive charge and make the atoms, as a whole, neutral. When too few or too many electrons are present, the atom becomes charged and receives the special name—ion. When an electron is removed from a hydrogen atom, an isolated proton remains—a hydrogen ion. Hydrogen ions can occur in water solutions (we recognize such solutions as acidic), as well as in space and inside stars.

Neutrons, unlike protons, cannot exist for long as isolated objects. A neutron set free will self-destruct in a short time (about ten minutes on the average) into a proton, electron, and another subatomic particle called an antineutrino. Only when they are under the influence of certain forces exerted by other nearby neutrons or protons, can neutrons remain stable indefinitely. We know of two circumstances when this happens. One is in atomic nuclei, as we have just seen. The other is when a whole starful of neutrons is bound together by their mutual gravity into an object fittingly called a neutron star. We shall encounter one in Chapter 11.

Actually, neutrons and protons themselves have a fairly complex structure that physicists have been studying in recent years. But we will not need to know about that structure in order to study life in the Universe. Therefore, we will not descend to yet deeper levels where this structure would become apparent. Instead we reenter COSMEL. We leave the —15 level and, with a single push of the G button, return to our kitchen. We can then take a final nostalgic look at the sugar and salt crystals, reduced to mere specks, and at the drop of soup, which a short time ago seemed a vast inhabited ocean.

Fable 1

The meeting of the Council of Microbes was very acrimonious on that long-ago day. The representative of the Amalgamated Bacteria held the floor and was speaking with great indignation.

"We must take immediate steps to end this pollution which is threatening our life-style, and even our very lives," he thundered. "If it goes on, it could render most of the Earth uninhabitable." He continued to make these points with various rhetorical flourishes, interminably.

When the delegate of the Blue-Green Algae at last was able to reply, he was equally irate. "First," he stated with great sarcasm, "let me speak that unmentionable word which my distinguished Bacterial colleague has not deigned to use —photosynthesis. It may be all right for *him* to speak of it as if it were some evil vice that could be given up with enough willpower, but *our* people need it to survive now, and the rest of you will depend on it for the future. We need to develop such alternatives before our resources run out. Besides, I have learned from a reliable source that certain members of the Bacterial nation have also been experimenting with photosynthesis."

The Algae delegate had not quite finished, but the head Bacterium had become so agitated that he could not be denied the floor. In fact, the alert efforts of his assistants were barely sufficient to prevent him from attacking the previous speaker.

"There is more than enough chemical food on this planet to take care of us!" the Bacterium shouted, "and our descendants, and even the Blue-Green Algae. Our proven reserves will last for thousands of generations! They will last even longer if we check our wasteful and selfish tendencies toward unlimited growth and live in a proper, balanced life-style. This way has been good enough for our ancestors for a billion years, and it should be good enough for us now.

"That last slanderous comment by my ill-mannered opponent is an attempt to confuse the issue," the head Bacterium continued. "True, some of us have been playing with solar energy, but in a restrained, ecological way, with no polluting side effects. Photosynthesis becomes dangerous when you attempt to use water as fuel. Although there has been an elaborate cover-up attempt, I know it as a fact that quite a few of the Blue-Green Algae have already died in the experiments. They perished because of the lethal by-product that their process produces—oxygen gas."

The very mention of that word caused panic in the entire Council. At that very moment, a chance breeze brought a few wisps of oxygen into the meeting area. The representative of the Methanogens, a minor group in the Council, became hysterical.

"Oxygen gas! We'll all die! We must flee at once to a safe place." He and his entire delegation started to rush off.

Both the head Bacterium and the Algae delegate tried to calm them. "It's not so bad," the Algae representative said, "we'll get used to it in a while."

But the Methanogen continued to cry out, "We'll be killed!"

"Where will you go to then?" asked the Bacterium.

"Into the mud," was the reply, "deep in the mud. It's the only safe place. Come and join us."

"That's no way to live," the head Bacterium shrugged. "We were born up here, and we'll stay up here, live or die." A few of his group rushed off after the Methanogens, but most of them remained in place.

When the uproar had subsided, the Council continued to debate, but it never did come to an agreement on how to control photosynthesis. As they argued, the levels of oxygen in the area continued to rise.

Chapter 3

Earth's Biosphere
and What It Eats

We do not get a complete picture of Earthlife by looking only at individual living things; we must also look at the individuals as part of a whole. The characteristic activities of Earthlife, such as the copying of nucleic acids and synthesis of proteins, are carried out in such diverse environments as fresh-water ponds, salt lakes, deserts, and cold Antarctic valleys. Active life can be found everywhere from mountaintops down to deep ocean trenches ten thousand meters below sea level, at a pressure a thousand times greater than on the surface.

A sizable part of the Earth's surface takes part in these activities. If we count the material contained at any one time in individual living creatures, it comes to several trillion tons, perhaps fifty thousand times as much as the weight of the whole human race, or the equivalent of 10^{30} bacteria. But even this large amount of living material is not everything that is involved in Earthlife. Much of Earth's atmosphere, oceans, and some of the soil have in the past and will again in the future be incorporated in individual living creatures. They are as much a part of Earthlife as an unoccupied chair is a part of the furniture in a house.

The sum of the once, now, and future living things of Earth has been called the biosphere. This term has been used in many ways. Here we will use it to include present living things as well as the part of the nonliving realm that enters into their metabolic activity. The parts of the Earth's surface that are not included in the biosphere are most of the rocks, some inert gases in the atmosphere, and perhaps portions of the permanent polar ice cap. By this definition, the biosphere forms a much greater proportion of Earth than just living things, and contains perhaps 10^{24} grams of material. An indication of the contents of the biosphere is given in Table 3-1.

At any instant it is an easy task to divide the biosphere into its living and nonliving components. Over longer periods, however, the substances in the biosphere are cycled and divide their time between residence in active life and in the environment. A carbon atom, for example, might spend years as part of a sugar residue in the DNA of a mammal. Upon the death of the creature, this residue would be released from the cell by bacterial decay. Assimilated by the bacterium, the sugar fragment could eventually be used for fuel in respiration. The carbon atom would be released into the atmosphere as carbon dioxide. After residing in the atmosphere for hundreds of years, it might be fixed by photosynthesis by a plant on the opposite side of the planet. The plant could incorporate the carbon into a portion of a leaf, which might be eaten by an insect. The carbon atom could pass from predator to predator up the food chain, eventually reappearing in the DNA of a mammal. An alternative destination for the carbon (when bound into carbon dioxide in the atmosphere) would be the sea, where it would be converted to a form called carbonate. The carbon atom could eventually be deposited as limestone and remain within the stone for millions of years until it was liberated by erosion. It might then again take part in the active life of some organism.

Not only carbon but other elements cycle through the biosphere in the same way. The individual atoms retain their identity, being neither created nor destroyed in the chemical processes of the biosphere. But as they travel from living to

Table 3-1

THE CONTENTS OF THE BIOSPHERE

Parts of the Biosphere	Contents	Fraction of Biosphere's Mass
Atmosphere	Nitrogen	3×10^{-3}
	Oxygen	10^{-3}
	Carbon dioxide	10^{-6}
	Water vapor	10^{-5}
Condensed water	Salt water	.95
	Fresh water	.01
	Ice	.02
Soil materials used by life	Carbon	10^{-5}
	Other elements	10^{-5}
Active life	Sea plant life	10^{-7}
	Sea animal life	10^{-8}
	Land plant life	10^{-6}
	Land animal life	10^{-7}

Only a small part of the biosphere is active life, but the rest is cycled through the active life at various rates. The total mass of the biosphere is approximately 10^{24} grams, most of which is ocean water.

nonliving matter, or from one organism to another, their molecular form changes. We will see how green plants produce molecular oxygen by photosynthesis, and how animals bind the oxygen into carbon dioxide by respiration. This is another example of a cycle of the biosphere. Each molecule of oxygen in the atmosphere forms part of a living thing on the average of one day every ten years. Each hydrogen atom in the water of the ocean spends an average of almost one minute of each year in a living organism.

These cycles unite the biosphere as an independent whole, in which the functions of one segment are required in order to allow the functioning of another segment. The billions of years of evolution and adaptation have brought the biosphere into a state in which the activities of both the living and nonliving parts allow these cycles to work smoothly.

For example, the nitrogen in the atmosphere is not very reactive chemically under Earth conditions, and is not readily brought into combination with other atoms to form chemical compounds. Most living organisms in the biosphere are not able to carry out this process, known as the "fixing" of nitrogen. Some nitrogen is fixed by lightning acting directly on the molecules in the atmosphere. Nitrogen is an essential component of protein, and if the only supply of it in useful forms were the amount fixed by lightning, the biosphere would be much more limited in its extent than it is today. This fate is avoided because a few dozen types of microorganisms have learned to fix nitrogen. Some of them live in symbiosis within the roots of plants and furnish nitrogen to them, and eventually, as these are eaten, to animals. If these activities were the whole story, nitrogen would disappear from the atmosphere in a time that would be very short compared with Earth's history. Other microorganisms exist, however, that act on decaying organic matter and release some of the nitrogen into the atmosphere so that the total amount present in it changes slowly, if at all.

There are also cycles within the biosphere that seem to have little to do with living things. An example is the water cycle, which involves rain followed by runoff into rivers. The rivers flow into the ocean, from which water evaporates and eventually falls again as rain. Such cycles would occur even without Earthlife, and in some instances there has been a mutual adaptation between Earthlife and these preexistent cycles.

The pattern of cycles within the biosphere is an interesting parallel to the complex biochemical cycles that go on within individual cells and organisms. We have seen that the metabolism of a cell involves a long series of chemical reactions that mutually influence one another. They are arranged so that a product of one reaction is an essential component of another reaction. It is this interrelated set of processes that constitutes the life of a cell. In complex multicellular organisms as well, there are cycles in which one part of the organism influences other parts through hormones or through nerve impulses. These processes in turn are influenced by yet other parts in a way that maintains the organism over a long period of time. In a sense, the biosphere is to its individual parts as a complex organism is to the cells of which it is composed. Just as the cells do not "know" that they are part of an organism but are nevertheless highly constrained in their behavior by this fact, so all the myriad living things of Earth do not know that they are part of a superorganism—the biosphere—but that fact nevertheless influences their function and even the fact of their existence in countless ways.

While individual organisms live and die, and even whole species wax and wane over the long history of Earthlife, the biosphere has had a continuous existence for billions of years, starting even before the first recognizable living things arose on Earth. If the evolution of Earthlife has had any consistent theme, it has been the extension of the biosphere to ever-larger parts of the Earth, and the increase in complexity of the biosphere within its domain.

ENERGY, THE REAL STAFF OF LIFE

There is one process in the biosphere which remains to be discussed that in some ways is the most important process of all. For each of the cycles mentioned, the biosphere is self-sufficient. Different types of atoms are transferred from place to place, from living to nonliving parts of the biosphere and back, but few atoms are ever lost or added from outside. There is another property of matter known as energy, which is also transferred from place to place within the biosphere but in a cycle that is not closed. The biosphere needs a constant supply of energy in the form of sunlight from the outside, or its energy supply runs out. If that happened, then all the other cycles in the biosphere would stop and it would die. In order to understand how the biosphere works, and indeed to understand Earthlife, we need to know something about what energy is, the laws that govern its behavior, and its role in the functions of living things.

Energy is a property of matter that comes in several forms. When matter changes its motion, location, or its chemical or physical state, one form of energy usually is converted into another. In these conversions the total amount of energy remains the same. The different forms of energy can be thought of as analogous to different forms of money: i.e., bills, coins, checks, etc. In a bank different types of money can be converted, but always in a fixed ratio; for example, one green paper with Washington's picture can be converted to one hundred copper disks with Lincoln's picture. This conservation of money is a direct analogue of the physical law called the conservation of energy. In the case of energy, unlike that of money, nature does not allow a Bureau of Engraving and Printing to create more energy out of nothing.

Two forms of energy are important in our discussion. The simplest form is the kind that is associated with the motion of objects, which physicists call kinetic energy. A running man has kinetic energy, as does a planet orbiting the Sun. The amount of kinetic energy depends in a precise way on the amount of matter and how it moves; the faster the motion or the more matter involved, the greater the kinetic energy. The motion can be orderly, as in the example of the running man. However, it can also be disorderly and still be associated with kinetic energy. A familiar form of kinetic energy is heat, which is often produced from other types of energy. Heat energy involves a disorganized motion of the individual atoms in an object. The amount of heat energy depends on the temperature of the object. The higher the temperature, the faster the atoms are moving and the greater the energy content. We will see that heat energy cannot always be converted into other forms of energy, and so is not always useful for the purposes of life, many of which require these other forms.

Another type of energy associated with motion is called radiation. Most of the radiation we will consider is related to moving patterns of electric and magnetic forces and is known as electromagnetic radiation. Visible light, radio waves, and X-rays are all examples of this type. The various forms of electromagnetic radiation differ from one another through a property called wavelength, which for visible light

Table 3-2

SOME TYPES OF ELECTROMAGNETIC RADIATION

Type of Radiation	Typical Wavelength (Centimeters)	Number of Photons per Calorie
Radio waves	20,000	4×10^{27}
Microwaves	20	4×10^{24}
Infrared light	10^{-2}	2×10^{21}
Visible light	5×10^{-5}	10^{19}
Ultraviolet light	10^{-5}	2×10^{18}
X-rays	10^{-8}	2×10^{15}

The shorter the wavelength of a type of radiation, the greater the energy of each photon. As we proceed down the table, wavelengths grow shorter, and fewer photons are needed to add up to a total of one calorie.

determines its color. The types of radiation also differ to some extent in the way they interact with matter. Physicists have discovered that all electromagnetic radiation is made up of a swarm of small packets of energy called photons. For radiation of a specific wavelength, such as green light, the photons all have the same amount of energy. Radiation of shorter wavelength contains photons of higher energy. (Table 3-2.) The amount of energy in a sample of radiation depends on both the number of photons present and the energy of each photon, just as the amount of money in your wallet depends both on the number of bills and the denomination of each bill. However, the two factors usually are not interchangeable. Ten photons, each with one unit of energy, do not behave in the same way as one photon with ten units of energy. So, to fully describe the energy content of some radiation, we must know both the number of photons and the energy of each photon.

The other important form of energy is associated with the various forces that objects exert on one another, and is called potential energy. Potential energy depends not on motion but on the distance objects are apart from each other. A stone sitting on top of a mountain high above the ground has more potential for doing damage if it falls than the same stone atop a small bump on the ground. The term potential energy signifies that it can be converted to and from kinetic energy freely. As an object falls toward the ground, its potential energy decreases, while its speed and kinetic energy increase. Conversely, an object thrown upward is gaining potential energy as it loses speed and kinetic energy.

Potential energy can also be converted into heat. A star is formed when a giant cloud of gas in space begins to contract because of the mutual gravitational attraction of the atoms in the gas. As it contracts, the potential energy decreases and the gas heats up. Some of the energy also is emitted as radiation during the contraction process. Eventually, other processes begin to occur in the now-dense hot gas ball, and a new star is born.

Potential energy is of many types, depending on what forces act between the objects. The examples we have given are all gravitational potential energy. This is the most important form for large objects, but not for small ones such as atoms. A somewhat similar form is electrical potential energy, which is associated with the electrical attractions and repulsions between charged objects such as electrons and protons. This electrical potential energy is the dominant form in atoms and molecules, and even in objects as large as a stone. In each case it is this energy that provides the fundamental "glue" that holds the object together.

Many of the changes involving energy that take place in nature can be simply understood in terms of these forms of energy. For example, we will soon see that in many of the chemical reactions involved in Earthlife, energy in the form of heat or radiation is either produced by the reaction or must be supplied for the reaction to occur. One way to describe this is by saying that the reacting molecules contain chemical energy, while the product molecules contain a different amount of chemical energy, and therefore extra energy must either be supplied or removed in order for the energy accounts to be balanced. For example, if two molecules combine, one containing seven units of energy, the other five units, and produce a molecule containing only ten units of energy, then two units will be released as kinetic energy (heat) of the product, or sometimes as radiation. In any chemical or other process that takes place, this type of energy balance must be exactly satisfied.

This description, while accurate enough, is not complete. Chemical energy is really not a new form of energy. The chemical energy of a molecule is just the sum of the kinetic and potential energies of all the constituents of the molecule. When the atoms in two molecules recombine into a different form, as in a chemical reaction, the kinetic and potential energies change in a complex way. When we add up all the kinetic and potential energies of the components of the reacting molecules and compare this sum with the corresponding sum for the product molecules, the two will not in general agree. The difference must either be supplied from some external source, or will appear as excess energy, depending on whether it is negative or positive.

From an understanding of the detailed structure of atoms and molecules, it is often possible to know which chemical reactions will release excess energy and which will require energy from the outside. If not, this can be determined by measurements. A very important difference between the reactions that release energy and those that require energy is that the first kind can occur spontaneously when the reactants come together, while the latter must rely on the extra energy being present also. It is like the difference between depositing money in a bank, which can almost always be done, and borrowing money from a bank, which requires the bank's agreement. We will now see that both types of reaction play important roles in Earthlife processes.

WHAT MAKES LIFE RUN?

Instead of directly discussing the complex role of energy in Earthlife, let us first examine a simpler case—the workings of an automobile. Imagine that a child is inspecting an automobile in the same way we examined our bacterium in Chapter 2. He opens the trunk, peers under the hood, sits in the driver's seat, and crawls beneath the auto to inspect the tires and axles. In the end he might still ask, "What makes it run?"

In answering, we would have to explain the need for fuel. We would indicate the place within the engine where the fuel is burned. We would then demonstrate how the small explosions which result are used to drive pistons within a cylinder. This energy is ultimately harnessed to turn the wheels of the car.

Our experience on the —8 level of COSMEL has prepared us to understand the burning process in terms of the molecules involved. If we take a trip into the gas tank, we would find that gasoline is a mixture of molecules containing just two types of atom; we would recognize the black and white of carbon and hydrogen. The name applied to molecules containing just hydrogen and carbon is, not surprisingly, hydrocarbon. In the burning process these molecules combine with oxygen from the air to produce carbon dioxide and water. We can summarize this in a simple chemical equation:

$$(C, H) \quad + \quad O_2 \quad \rightarrow \quad CO_2 \quad + \quad H_2O \quad + \text{ energy}$$

(C, H)		O$_2$		CO$_2$	H$_2$O	
gasoline		oxygen		carbon	water	
(a mixture)				dioxide		

Evidently, this is an example of the type of reaction in which extra energy is left after the molecules have recombined.

If the burning process is not carried out properly, insufficient oxygen may be present. In that case many of the carbon atoms may have to make do with only one oxygen atom, producing carbon monoxide (CO), or with none at all, yielding soot. (Carbon monoxide is, in fact, formed in automobile traffic. A deadly gas, it is one of the hazards of modern urban life.)

Living things on Earth, like automobiles, require energy for their activities. There are proteins to be assembled, cell membranes to be constructed, flagella, cilia, and arms to be waved. Some of these activities are chemical reactions that need a supply of energy to make them go. Others involve motion and so involve the transformation of other forms of energy into kinetic energy. One important source of this energy is a process similar to that used in an automobile. A suitable fuel is combined with oxygen to furnish energy. The general name for this process is oxidation (the term is sometimes extended to cover processes where a related agent of combustion is substituted for oxygen). Most living things on Earth have difficulty in utilizing hydrocarbons, such as those in gasoline, as fuel. Sugars are preferred, even though less energy is available from their combustion. (The presence of oxygen atoms in sugars indicates that they have already been partly "burned.") This energy is still substantial, however. It is not released all at once, as in the piston of a car, but in gradual stages. It is stored in other types of chemical bonds and used as needed. We have already encountered one of the most widely used of these energy banks—a nucleoside triphosphate, a molecule called ATP. As you may recall, it was a subunit used for the construction of messenger RNA. Removal of the extra phosphates not only prepares the subunit for assembly into RNA, but liberates chemical energy to make the reaction possible. The energy stored within the phosphates of ATP is also used in all living things on Earth to drive a variety of reactions which have nothing to do with the construction of RNA.

The process that living things use to get energy by oxidation of food is called respiration. We can summarize it as:

$$(C, H, O) + O_2 \xrightarrow{\text{respiration}} H_2O + CO_2 + \text{energy}$$

$$\text{sugars} \quad \text{oxygen} \qquad\qquad \text{water} \quad \text{carbon dioxide}$$

Sugar is not the only fuel that can be burned to obtain energy. Many other compounds containing carbon, hydrogen, and other elements can be used. When our bodies are given a large excess of chemical energy, we store it as fat, a substance containing carbon, hydrogen, and relatively little oxygen. Fat can be used as fuel when it is needed. Because it is less oxidized than sugar, fat contains three times as much energy per unit weight. This is why fatty foods contain more calories than nonfatty ones.

The simplest living things on Earth, such as bacteria, show the greatest versatility in the use of fuels. Strains of bacteria exist that can utilize almost anything containing carbon that can be burned. Even combustible mineral substances like sulfur and forms of iron are used by some species. Furthermore, some bacteria can replace oxygen with minerals containing oxygen, as the oxidizing agent. Many variations on the respiration theme therefore exist.

Respiration itself can be dispensed with in favor of other reactions that release chemical energy. The alternatives are generally less efficient in energy release than respiration. Most living creatures on Earth, including ourselves, have at their disposal a process called fermentation. It is useful to us as a temporary expedient when insufficient oxygen is available in our muscles to burn sugar, as in a long-distance race. In fermentation the sugar molecule is broken in two, freeing a modest amount of energy. When oxygen becomes available again, the remainder of the chemical energy originally in the sugar can be extracted.

One living thing is noteworthy for the manner in which it extracts energy from chemicals. This is a type of bacterium known as a methanogen. It is killed by oxygen and survives only in special oxygen-free areas such as the muddy bottoms of the Black Sea and San Francisco Bay. Hydrogen gas is produced in those environments by decaying animal and plant remains. Hydrogen is the lightest gas; it has been used in the past to inflate balloons and dirigibles. We did not encounter it among the molecules of air on the —8 level. One reason for this absence is also why its use in dirigibles has been discontinued. Hydrogen tends to react with oxygen, sometimes explosively, to form water. An environment may contain a lot of hydrogen, or a lot of oxygen, but not both at the same time, at least not at Earth temperature.

The methanogens obtain energy by combining hydrogen with carbon dioxide to form methane gas and water. Methane, containing one carbon and four hydrogens, is the simplest member of the hydrocarbon family. A process such as this one, in which a substance is combined with hydrogen, is called "reduction." Very often the chemical change in a carbon-containing compound produced by reduction is the reverse of that produced by oxidation.

Our description of the energy supply for Earthlife is far from complete. The processes described thus far would suffice if supplies of chemical energy were inexhaustible, and sinks existed in which infinite amounts of waste products such as carbon dioxide could be dumped. This is not the case, neither today nor in the history of life on Earth. Green plants play an indispensable role on Earth by using sunlight in the process called photosynthesis to replenish the supply of chemical energy. This process, like respiration, is complex and involves many steps. Its overall effect can be summarized by reversing the equation given on page 86 for respiration:

$$\underset{\substack{\text{carbon} \\ \text{dioxide}}}{CO_2} + \underset{\text{water}}{H_2O} + \text{energy} \xrightarrow{\text{photosynthesis}} \underset{\text{oxygen}}{O_2} + \underset{\text{sugars}}{(C, H, O)}$$

This equation describes the most important form of photosynthesis. Variations do exist, with the simplest organisms again displaying the most diversity.

Both of the substances used in respiration are the products of photosynthesis. The sugar produced by plants is used either by the plants themselves or by the animals that eat them. The oxygen which combines with sugar in respiration has been liberated by photosynthesis, perhaps in the distant past. The energy released in respiration is derived from the energy in the sunlight that was absorbed during photosynthesis. This energy was stored as chemical energy in sugar and oxygen. Chemical energy is more readily stored and used than sunlight. It is available even when the sun is not shining. Living plants and algae long ago solved a problem that we humans are now grappling with as we attempt to use solar energy to run our civilization.

WHY DOES LIFE NEED AN ENERGY SUPPLY?

The activities of Earthlife transfer energy from place to place on Earth, and within individual living things, in an endless dance. But why must energy be constantly supplied from an outside source, the Sun? Since the total energy does not change in any reaction, why cannot Earthlife simply maintain its energy content by careful husbandry as it has learned to do with matter?

In answering this question, let us imagine that a series of energy conversions is analogous to a poker game with many players. At the end of each hand a portion of the pot is put into the kitty. If the game runs long enough in this fashion, virtually all the money will end up in the kitty, and the players will have none. If more money is continually added to the game from outside, however, it can continue indefinitely.

Transformations that involve energy operate in a similar way. A portion of the energy in each exchange becomes unavailable for further conversions, like the money in the kitty. The energy that can still be used (the money still in the hands of the poker players) is called free energy. Radiation, energy of orderly motion, and chemical energy are forms which remain generally available, while heat energy is more restricted and a portion of it represents the energy that is unavailable (the money in the kitty). The law which governs the conversion of available into unavailable energy and determines what part of heat energy remains available is fundamental in physics. It is called the second law of thermodynamics.

If living things on Earth tried to maintain themselves on a fixed amount of energy, they would suffer the same fate as the poker game in its original form. Their life would come to an end as the energy was all gradually turned into useless heat. Individual living things, and Earthlife as a whole, avoid this fate by continually taking in free energy from the outside. Some of the energy is used for the purposes of life, while the remainder is returned to the environment as heat. Living things elsewhere in the Universe must also behave in this manner, as the second law of thermodynamics applies universally. In a later chapter we will discuss this behavior in greater detail, because it is a central feature of our definition of life. But first we wish to continue our exploration of Earthlife and consider the important question of how it first arose on this planet.

Chapter 4

How Earthlife Evolved

Earthlife as it exists now is very complex. This complexity of organization is true both of the biosphere as a whole and its individual components. Even the simplest known bacteria have a highly organized structure, more complicated than such human institutions as a legislature or a factory. They are certainly much more intricate than the nonliving components of our environment. This complexity of the biosphere is the product of a process of evolution that has taken a very long time to occur, by human standards. The study of Earthlife in its present form is fascinating in itself, but it gives us little clue about the likelihood that other biospheres will develop elsewhere in the Universe. In order to determine this, we must follow a faint trail of circumstantial evidence back through time, and infer the nature of the steps that led to the emergence of the primitive biosphere from its environment. The character of that environment itself is another crucial detail, and to understand what we can of it, we must follow our quest back beyond the creation of the Earth to the birth of the Universe itself.

TIME AND THE UNIVERSE

The Universe as a whole appears to be about ten to twenty billion years old. Back then, all the matter and energy in the Universe were in a very different form than at present. They were compressed into a hot, dense mass, and the very atoms that exist now were not yet made. Only their components existed, in the form of many subatomic particles. What came before this stage is not clearly known yet and not relevant for our purposes. Although there may have been some complex forms of matter in this early stage, it is extremely unlikely that they could have survived into the present phase, in which matter and energy are much more rarefied. Conversely, any forms of life that could thrive today would almost certainly have found the very early stages of the Universe uncongenial and so could not have developed. We can therefore in good conscience begin our story of life with the "Big Bang," from which the primeval ball of matter and energy started to expand.

Ten billion years is a long time, but actually many familiar things are almost as old. The Sun and Earth have been here for about five billion years. The Earth, like Burgon's city of Petra, is "half as old as Time." The hydrogen atoms that are present in all the water on Earth are even older, and date back to only a million years after the Big Bang. According to present ideas, most of the other chemical elements were produced at a somewhat later stage of the Universe. This occurred after stars had formed and passed through the parts of their life cycles in which these elements were made. At any event, the history of our planet goes back over four and a half billion years. This amount of time is immense and hard to relate to in our daily lives. To gain perspective, we will start with an overview of the entire process, in which we have condensed the history of the Earth into a week. In another well-known account of the creation, the job was done in only six days, with a subsequent day set aside for rest. Our understanding of the events that took place, particularly the crucial early ones, is incomplete, and many areas of disagreement and gaps in our knowledge still exist. When we conclude our overview, we will explore the controversies and the nature of the evidence that exists in more detail.

Table 4-1

EARTHWEEK*

Time During Earthweek	**Actual Time Before Present**		**Important Event**
Sunday after midnight	4.5	billion years	Earth is formed
Early Sunday evening	4	billion years	Earth's crust solidifies
Late Monday evening	3.4	billion years	Earliest *known* life
Early Tuesday morning	3	billion years	Bacteria and algae exist
Wednesday noon	2	billion years	Sizable amounts of oxygen in atmosphere
Thursday evening	1.2	billion years	First eucaryotic cells
Friday afternoon	900	million years	First multicelled organisms
Saturday morning	600	million years	First land animals
Saturday afternoon	200	million years	First mammals
11:50 Saturday evening	4	million years	Earliest hominids
11:59 + 59 seconds Saturday evening	5	thousand years	Recorded human history begins

* A summary of the history of the Earth reduced in time scale to one week.

THE DEVELOPMENT OF LIFE ON EARTH: FOUR AND A HALF BILLION YEARS IN A WEEK

As our tale begins just after the birth of the Sun, it is fitting to pick the first instant after midnight on Sunday as the starting point of our week. Scientists are not sure whether the Earth was completely formed when the Sun began to emit energy from the nuclear fusion taking place deep within it. If not, the Earth was formed soon thereafter, sometime on Sunday. The oldest rocks on Earth that we know of are some 3.7 billion years old, about 75 percent of the age of the solar system. So whenever the Earth was formed, it was solidified by late Sunday evening, or perhaps much earlier in the day.

The conditions on the newly solidified Earth remain somewhat mysterious. Were there oceans and continents as there are at present? Was there an atmosphere, and if so, what did it contain? There has been much speculative reasoning about this, but little certainty. A picture accepted by many scientists is that not long after the Earth's crust solidified, there were large bodies of water on the surface, but no continental-sized bodies of land. The continents developed gradually over the week of Earth's history. There was also an atmosphere of a very different kind from the oxygen-nitrogen mixture of the present day. The primitive atmosphere, which emerged gradually from the interior of the Earth, consisted of gases such as methane, nitrogen, hydrogen, water vapor and carbon dioxide, but no free oxygen. This combination was reducing rather than oxidizing in character. Equally important for the development of Earthlife was that some of the gases present could absorb ultraviolet light from the Sun and then undergo chemical reactions to produce more complex molecules. Some of these molecules were heavy enough to fall onto the oceans' surfaces and dissolve. Over a period of time, large amounts of substances containing carbon, nitrogen, oxygen, and the other atoms in the original atmosphere, gradually were absorbed into the early oceans. Some of the energy derived from sunlight remained stored in the chemical energy of these molecules and was readily available for the first stages of life.

We do not know the precise mixture of the complex molecules formed in this way. Attempts have been made to simulate this process of formation in laboratory experiments, by passing ultraviolet light or other sources of energy through mixtures of some of the gases that are thought to have existed in the early atmosphere. In these simulations a variety of molecules containing carbon are formed, including some amino acids that play vital roles in Earthlife. But also many other molecules are formed in the experiments that play no role in Earthlife. The amounts produced depend very much on the exact conditions of the experiment. Since the experiments are done under rather different physical conditions than those in the real situation, it is doubtful that any firm conclusions about what was present in the early oceans could be drawn from these results. The best that we can say is that it is possible to form molecules containing from ten to twenty atoms, including some carbon atoms, by the action of sunlight on reducing gases, so that no special creation is needed to bring the story of Earthlife up to that point.

Almost certainly there was a period of chemical evolution before the first individual living things evolved. During this chemical evolution, fairly complex molecules were able to influence each other's formation in a way that led to a gradual increase of some types of molecules at the expense of others. We can imagine simple ways for this to happen. For example, if there are three molecules A, B, C, such that the presence of B aids in the formation of A from simple chemicals, the presence of B inhibits the formation of C, and the presence of A helps the formation of B, then there will be a tendency for A and B to increase in quantity, while the quantity of C remains small. Processes of this type, and similar, more complex changes, would lead to a concentration of certain molecules in the primeval oceans and a depletion of others, compared with the random amounts produced by the action of light on the atmosphere.

Probably Earthlife developed through the reactions of these carbon-containing molecules in water over a period of time. But we have no direct evidence of this nor any real understanding of the process. We don't even know how long it took. The oldest discovered traces of life extend back almost as far as the oldest rocks, to Monday evening of our standard week (3.4 billion years ago). It is possible that life is even older and appeared shortly after the Earth solidified.

It is difficult to determine an exact date, because the principal source of evidence that we have to rely upon is fossil remains. However, few specimens have survived from such ancient times, and those which remain are often distinguished only with difficulty from irregular structures normally occurring in rocks. Frequently mistakes are made. To cite evolutionist G. G. Simpson, "Eozoon, proudly named the 'dawn animal,' is now considered to be no animal at all nor yet a plant or any form of life but a mere inorganic precipitate." We see that it is difficult to distinguish the living from the nonliving on the basis of appearance.

It was once thought that a long time was needed for life to develop through the action of chance on the combination and recombination of atoms. But we will see that even the long history of the Earth is an instant compared with the near-eternity that the chance reshuffling of molecules would take to produce the simplest living form. Something else is required, which has not yet been clearly understood. When scientists clarify how the process of prebiotic evolution occurred, we will be able to estimate how long it took for life to develop. Meanwhile the best we can do is search for earlier and earlier fossil-like remains to help us resolve the questions by observation.

Over three billion years ago, early on Tuesday morning of Earthweek, individual living things similar to present-day bacteria and algae (microscopic plants) already existed. These individuals already had the capacity to reproduce themselves. They accomplished this by containing, within a somewhat porous membrane, two sorts of molecules—nucleic acids and proteins—already described in Chapter 2. The ability to reproduce requires both types of molecules in present-day organisms. Conceivably, primitive things were able to reproduce only through nucleic acids, but there is little evi-

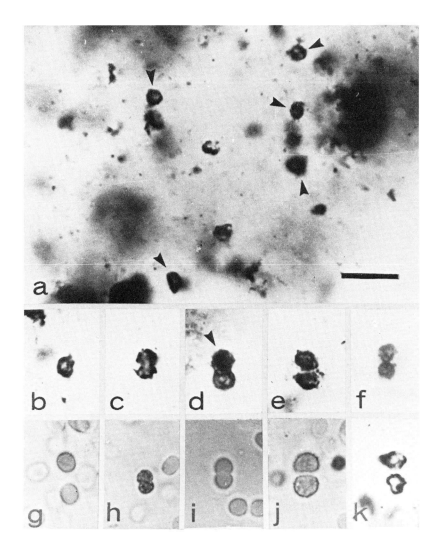

Plate 7. Microscopic fossils of one-celled organisms observed in South African rocks that were formed over three billion years ago. The arrows indicate individual cells. Inserts b through k indicate cells that appear to be in the process of division.

dence for this. Some scientists, such as Richard Dawkins in his book *The Selfish Gene*, have identified the origin of life with the first molecules that could reproduce themselves. We think that this view is misleading, for a reason that will be discussed in more detail later. Life originated not in a single step, but as part of a long evolutionary process in which the appearance of nucleic acid or other replicating molecules was only one step in the middle of the evolution. But it was an important step, and, once invented, replication was a strategy that Earthlife has used ever since.

One reason is that the replication of nucleic acids was a convenient method for evolution to occur. A particular arrangement of bases along the nucleic acid chain that offers some advantage was replicated more often than others, and that arrangement tended to become the dominant one. If, because of some random change in the arrangement, a still more advantageous strand was produced, it was replicated even more often and in turn became dominant. This process continued until eventually the molecules and processes involved in the replication of nucleic acids became the central feature of Earth's biosphere.

The earliest living things probably made use of the large amounts of chemical energy that existed in the carbon compounds dissolved in the seas. In effect, they lived off stored sunlight, just as we do when we burn coal or oil. But this process could not go on for very long. If the number of bacteria on the early Earth were about the same as at present, they would use up all the available chemical energy in a comparatively short time, less than a million years.

Microorganisms might have avoided this for a while by limiting their numbers, but living things generally do not plan for their future in that way, as can be seen from the present human population explosion. Instead, they solved their "energy crisis" by the invention of photosynthesis. Some of them, the blue-green algae, learned to produce a substance that efficiently absorbs visible sunlight and transforms it into chemical energy. In this way they were able to tap a constant flow of energy for their own purposes, and multiply to much greater numbers than was possible when they used only stored chemical energy. Furthermore, other living things developed which fed on the algae, thus obtaining their energy indirectly from the Sun.

Eventually, forms of photosynthesis were developed that led to the production of free oxygen from water. This oxygen, produced in greater and greater quantities over the aeons, gradually led to a change in the atmosphere from an original reducing state to its present highly oxidizing state. This transformation had occurred by early Wednesday afternoon of Earthweek (about two billion years ago). The oxygen was probably lethal to many of the living things that existed when photosynthesis began, and only those which could coexist with oxygen were able to survive this byproduct of the algae's way of storing energy. A new lifestyle, which eventually led to the development of the animal kingdom, was also made possible by the presence of free oxygen. Organisms could ingest others, or their products, and combine their meals with oxygen to obtain energy. We described this process, called respiration, in Chapter 3.

The change in character of the Earth's atmosphere is perhaps the most spectacular effect which life has produced. It is an indication that in other environments, life may also have significant effects that we can someday identify.

The oxygen produced by photosynthesis had another effect, which was to eventually screen out from the Earth's surface most of the ultraviolet light that originally started the life process. It did this by creating a complex chain of chemical reactions in the atmosphere, eventually producing the acrid gas, ozone, a variant form of oxygen. Ozone absorbs ultraviolet light so strongly that a very thin layer of it in the upper atmosphere keeps much of the Sun's ultraviolet rays from reaching the surface. This effect allowed the land, and the top layer of the oceans, to become homes for life. Previously, living things which are easily destroyed by the ultraviolet light that ozone absorbs were probably confined to regions several meters below the surface of the oceans. Ultraviolet light does not penetrate those levels because it is absorbed by the water. This process took some time to happen, and the land areas probably became habitable only about a billion years ago, on late Friday morning of Earth-week. The oldest known remnants of land animals are only half as old as this, appearing about 5 A.M. on Saturday morning of our Earthweek.

One other step in the development of life happened along the way, the appearance of multicelled organisms. Scientists think that this happened about a billion years ago, also on late Friday morning of Earthweek, although some would put it much farther back. This happened in two stages. First, in some cells the part containing the genetic material developed into a separate unit, the nucleus. Various other structures, which we have seen in the cell on our trip with COSMEL, also developed. These cells evolved more complex mechanisms for cell division. All these changes led to a new type of cell called eucaryotic. Some single-celled organisms, such as paramecia, and all multicellular ones are eucaryotic. The simpler type of cell, present in bacterial and blue-green algae species, is called procaryotic. The development of eucaryotic cells apparently was soon followed by multicellular creatures. It is important to realize that life on this planet has spent about three-quarters of its existence in single-celled form, and even today the majority of organisms still exist as single cells. The evolutionary pressure to become complex is evidently not very great. Had we been able to inspect this planet for signs of life throughout most of its history, we would have detected nothing without the aid of a microscope. For that reason detecting life in other parts of the Universe may be more difficult than one might first guess.

Once multicellular organisms developed, the pace of further evolution quickened considerably. All the familiar forms of life, such as trees, fish, reptiles, and mammals, evolved within the last half-billion years, on the last day of Earthweek. The things we commonly associate with life are all relatively new. Human beings, by a generous estimate, originated two million years ago, or later than five minutes to midnight on Saturday evening of Earthweek. Recorded

100

history is less than five thousand years old, or within the last second of Earthweek.

Our account has now reached the present time. It is awesome to think of what events may occur in the next "week" on this planet. We will not speculate on this, but rather look at the mechanisms that have functioned to bring Earth's biosphere to its present evolved state. Other biospheres elsewhere in the Universe may be constructed of very different substances but still have similar mechanisms at their disposal.

We can divide the history of life on Earth into two periods, according to the extent of our knowledge concerning the processes involved. We have a good idea of how existing complex life forms developed from earlier, simpler ones. This knowledge is based on the great achievement of nineteenth-century biology, the Darwinian principle of natural selection.

NATURAL SELECTION

A number of processes, operating largely at random, introduce variability into the population of living things. Many of these variations lead to a decreased capacity for survival. These unfortunate organisms give rise to less offspring, and to none at all where the change causes the loss of some vital function. Organisms that have inherited a capacity to increase their chances of survival in the environment they inhabit have more descendants than average. Favorable traits are selectively propagated and are established in the realm of life.

The mechanisms which produce variations include mutation, genetic recombination by sexual means, and symbiosis. It has also been discovered recently that many genes of eucaryotes are not stored in one continuous length but are interrupted. It is as if unrelated messages were introduced at random into the sentences in this book. The purpose for this arrangement is not understood at this time. It may be part of another mechanism for obtaining genetic variability, since part of different genetic "sentences" can be connected to each other. The sexual mechanisms of heredity are familiar to us from our own lives, and were first put on a firm scientific basis in the nineteenth century by Gregor Mendel. They allow existing genetic information to be redistributed among the members of a population. Thus, two favorable features which developed independently may be combined in one individual, with even better effect on survival.

Our knowledge of the mechanism of mutation has developed only within the last quarter century. Mutations are changes in the data that is carried in the sequence of bases in DNA. They occur at random, creating information that has not existed before. Imagine a process in which you select a place in the text of *Hamlet* at random. You then decide arbitrarily whether one or more words are to be removed, added, or exchanged for new ones. If new words are to be added or used to replace existing ones, you select them, again at random from a dictionary. Usually this process will not increase the literary value of the play (try it yourself and see). In very unlucky cases, the meaning of an entire section may be damaged. In rare instances, however, the text will be improved. The play may have been optimal for its original audience, and at that time any changes made by the random process would have been for the worse. But changes in society and language since the play was written have made *Hamlet* increasingly less understandable to modern audiences. (This is analogous to a need for adaptation of an organism due to changes in its environment.) It is a tribute to the genius of its creator that *Hamlet* has retained so much meaning after the passage of centuries. One can see, however, that if the random replacement process were carried out a great many times and if the errors were discarded and improvements preserved, an updated *Hamlet* would eventually be produced.

The above analogy also illustrates the effect of random

mutations on living beings. A change in an amino acid in a protein that is coded by an altered segment of nucleic acid will occasionally lead to loss of function. Occasionally, an improved function will result. Once in a rare while, the new protein may have some completely new, desirable property. This possibility is more likely when the environment has changed in a way that puts new constraints on the organism. The process has a better chance of working when an organism has more than one copy of a gene—a common condition among higher organisms. In such cases an alteration that confers some new function upon a protein can be allowed, even if the old function was vital and was destroyed. The remaining unaffected copies of the gene will continue to produce the old form of the protein.

One consequence of natural selection is a pressure to develop species with the capacity to occupy new, previously uninhabited portions of the environment. At one point in history, the oceans may have been overcrowded with life. Organisms that first developed the ability to sustain themselves on land had a clear advantage. They were freed from the competition for the limited resources of the sea and could expand their numbers rapidly. A more recent, if less happy, example is the development of some bacteria which are resistant to antibiotics because they are capable of producing a substance that deactivates the antibiotic. While these bacteria would have no advantage in their usual environment, they have a considerable advantage in a patient who has ingested a dose of the antibiotic. As a result they will continue to thrive there, while their cousins fall to the ravages of the chemical. A further illustration of this effect may soon be provided by our own species. At the present stage in Earth's history, the capacity of our entire planet to support additional human life is being severely taxed. It is significant that, through technology, humans are developing the possibility of expansion into a place that Earthlife could not previously inhabit—interplanetary space.

Natural selection has been effective in bringing life from

its early beginnings through the development of intelligent beings living in society. It has also been inefficient and slow, taking perhaps 25 percent of the time since the creation of the Universe for the accomplishment of that task. In our earlier analogy, it probably occurred to you that more efficient methods exist for the conversion of *Hamlet* to modern English. A skilled editor with a thesaurus could probably do the task within months by the proper application of intelligence. Using the trial-and-error approach, many centuries would be required. Similarly, the application of intelligence can immeasurably speed up the process of evolution and harness it to the production of life forms needed to improve the quality of human existence. After four billion years of evolution, it is only within the last few millennia, with the introduction of plant and animal breeding, that this has been done. At present, powerful new methods for changing the genetic character of organisms are being developed. These methods, sometimes called genetic engineering, promise to radically improve our ability to influence biological processes for human purposes. We are indeed living at a unique moment in history.

A COMMON ANCESTOR OF EARTHLIFE

While natural selection can account for the apparent diversity of life that existed before and presently is found on Earth, it does not explain the reverse aspect of Earthlife, its remarkable biochemical unity.

We have seen that all known living cells share a large number of common features: They store genetic information in nucleic acids; they use proteins as catalysts and building materials; they store energy in carbohydrates; they construct their large molecules from only one of two possible mirror-image subunits; they assemble their proteins at ribosomes; they control the entry and exit of materials by cell membranes, and have many other processes in common. The most reasonable explanation for these similarities is that all living species on Earth today are descended from a common ancestor which possessed these features. This ancestor would itself be quite complex, the product of an evolutionary process. Were there then no differing species that shared the world with it, the products of divergent evolutionary processes? Radically different answers can be given to this question. The most obvious is that other species did exist but then became extinct (or perhaps, as we will discuss later, fled to remote and specialized habitats). This would be possible if one or more of the features in the common ancestor were a dramatic evolutionary improvement. If so, it would presumably outperform its competitors and sequester all the available organic material for its own purposes. Thereafter, no living thing not descended from this ancestor could evolve. The established line of life would evolve further and diversify to yield its present forms, but essential biochemical features would be maintained. It is also possible that a new life form caused fundamental changes in the environment so that previous types of life could not live anymore. The introduction of free oxygen into Earth's atmosphere by photosynthetic organisms is one example of this, although, as we have seen, it did not destroy all previous organisms.

EVOLUTION IN THE TEST TUBE

If there was a common ancestor of all Earthlife containing all the essential biochemical mechanisms we find today, it would still be a rather complex living thing, which could not arise by chance arrangements from nonliving matter. It is reasonable to suppose instead that the common ancestor itself evolved through natural selection from yet simpler living things. Proceeding backward, some scientists have arrived at the concept of the primitive replicator: the simplest living thing that can control its own reproduction and evolve by natural selection. This replicator might be little more than a macromolecule or complex of macromolecules. It would depend upon the environment to supply usable energy sources and its own components.

The possibility that Darwinian evolution through natural selection could proceed in a simple system composed of macromolecules has been demonstrated in a brilliant series of experiments by Dr. Sol Spiegelman and his colleagues. They used the simple virus, $Q\beta$, which has a single strand of 3,500 nucleotides of RNA as its genetic material. This amount of genetic material is enough to code for only a few proteins. Among those coded are the protein that forms the protective coat of the virus, and an enzyme called the replicase, which is needed for its reproduction. Like other viruses, $Q\beta$ depends upon its host (a bacterium) for ribosomes, for components of its nucleic acid and proteins, and for other necessities of life. Under simplified conditions, however, the viral RNA molecule can be made to reproduce without a living host. The necessary ingredients include the RNA, the four subunits from which it is assembled (nucleoside triphosphates), the replicase, a pinch of magnesium salt, and the proper temperature and acidity.

In this favorable environment, the RNA is first converted to a double-stranded helix. Each strand is then copied further to produce additional helices. This reproductive process can continue indefinitely, as long as living space and the supply of nucleoside triphosphates hold out. In order to avoid the necessity of growing the colony in a series of vessels of ever-increasing size (perhaps culminating in a swimming pool), a procedure called serial transfer was adopted. At intervals of about twenty minutes (almost enough time to double the colony size), a small portion of the colony is withdrawn and added to a fresh test tube containing the necessary ingredients for growth, including additional enzymes and triphosphates. Using this procedure, dozens of generations can be raised within a day in a convenient volume.

In one experiment the colony was taken through seventy-four transfers. During that time, the interval between transfers was gradually decreased. At the end of the experiment, the RNA present was analyzed. The major species was only one-sixth of the length of the starting viral RNA. Because of its shorter length, its generation time was only one-fifteenth that of the starting RNA. It is as if each generation of RNA were a letter to be typed and the enzyme were the typewriter. The shorter the message in the letter, the more rapidly it would be completed and the next letter started.

What had happened? Under the experimental conditions, the RNA was provided with very favorable conditions for its survival and growth. It was in a viral Garden of Eden. It had no need to create (code for) anything at all. The replicase, the subunits, all were provided. There was a need to maintain a certain critical length and message necessary for it to be recognized by the copying enzyme (it would be as if the typewriter refused to copy any letter that did not contain a certain key message). Under the conditions of serial transfer, an RNA molecule that could reproduce much faster than its fellows would dominate the population. If an RNA molecule suffered a random break but retained that portion necessary for recognition, natural selection worked in its favor, and it outcompeted the starting RNA. Any molecule, of course, which lost some of the critical region was not copied at all and was essentially "dead." Victory in this molecular competition went to the shortest viable piece that was formed by random processes.

Imagine that the typewriter would type any message containing the words "Type this message." If the original message was "Type this message over again," it would be typed. Suppose that a mistake was made, leaving out the word "again." The new instruction, "Type this message over," would still satisfy the typewriter's requirements, and it would get retyped; while if the first word, "type," was omitted, the resulting message would not trigger the typewriter. Eventually the message might get reduced to just enough to trigger the typewriter; that is, the phrase "Type this message."

The Spiegelman experiment had one feature that was atypical of Darwinian evolution. The system evolved in the direction of information loss, of lesser complexity. Later experiments had greater similarity to biological situations. In one case a drug was applied, which slowed down the rate of reproduction of the RNA. When serial transfers were carried out in its presence, a mutant RNA arose which was resistant to the added substance. The drug preferred to bind at a particular sequence of bases in the original RNA. The mutant RNA had undergone changes in its base sequence so that the favorite binding place of the drug was destroyed. Once again, the RNA had evolved in a direction in which its reproduction rate was increased. This shows how the earliest living things may have gradually increased their capacity to survive, in the face of various environmental challenges, by modifying their genetic information.

In addition to demonstrating natural selection at the molecular level, the above studies suggest the type of situation that may have existed when the first living things arose. We have the primitive replicating unit, the needed parts for its construction, a protein catalyst, and a source of chemical energy. In this particular case, the energy is present in the phosphate-to-phosphate bonds of the subunit. The decomposition of these bonds, releasing energy, is needed during the construction of each new RNA molecule. The process of RNA assembly involves the formation of water molecules and their release into a water solution. This requires energy and will not occur unless some energetically more favorable transaction takes place at the same time.

In the Spiegelman experiments, the old question about the chicken and the egg is reduced to a more fundamental form, "Which came first, the nucleic acid or the protein?" The nucleic acid cannot reproduce without the enzyme (protein). The enzyme cannot reproduce at all, but is needed for the nucleic acid to reproduce. The information needed to construct the enzyme is coded in the nucleic acid. It seems unlikely that two such complicated molecules could have evolved simultaneously and independently, yet neither can function as they do in living things without the other.

In order to resolve this dilemma, a number of experiments have been run. In one set, efforts were made to replicate a nucleic acid without an enzyme. Some interesting leads have turned up, but no conclusive demonstration has been achieved. Even less success has been encountered in the opposite approach. There is no evidence that a protein can carry genetic information and catalyze its own reproduction.

Some scientists have avoided the dilemma by focusing their attention upon the evolution of another critical cell component—the membrane. In the Spiegelman experiments, the test tube took over one of the crucial roles which the membrane fills in cell biology: that of keeping the components of the system in close proximity to one another. If the RNA, replicase, and nucleoside triphosphates had been washed down the kitchen sink, very little replication would have taken place in the plumbing!

Several efforts have been made to simulate the formation of primitive cells (protocells). Certain microdroplets called coacervates, which are readily formed from aggregates of large organic molecules, have attracted attention. Coacervates have boundaries and can concentrate nucleic acids and proteins in their interiors. Schemes have been proposed in which such protocells would grow by absorbing organic material and, after reaching a certain size, split by some mechanical process. It is not clear how such a system would convert itself to one that used a conventional genetic form of replication.

These various approaches to the origin of life unite on one fundamental point: the importance of establishing a primitive replicating unit. As we have noted, the replicator would first rely on the environment to meet its needs for energy and parts. As certain items fell into short supply, the replicator would evolve, by natural selection, to a state where it could aid in the manufacture of the scarce items. By gradual evolution, the various patterns characteristic of metabolism would be acquired, and something like primitive cells would be formed.

Although there is no direct evidence in favor of this scheme, and even the chemical identity of the original replicator is unknown, it is reasonable and fits in well with our knowledge of evolution at higher levels of organization. But a crucial problem still remains: explaining how the first replicator arose. It may have been composed of a protein and a nucleic acid within a primitive covering. Or perhaps, through some mechanism we do not understand, one type of complex macromolecule would suffice. As we shall see, there are gaps in understanding how even such a simplified replicator would arise from nonliving materials. In the next chapter, we will consider various solutions that have been proposed.

Fable 2

The members of the Little Puddlian Philosophical Society were engaged in a debate about the possibility of life outside of Little Puddle. The chief advocate of the unrivaled fitness of their own environment for life, a medium-sized male rotifer named Philo, was speaking. This he accomplished by wriggling three of his many cilia, producing sound waves in the water of Little Puddle.

"It is obvious that life is impossible outside the confines of Little Puddle. The environment we inhabit is optimally designed for life. Its water remains at about the same temperature at all times, whereas even a small change would be lethal to us. The balance of acidity and alkalinity is exactly right for living things because of the small amounts of nitrates and phosphates dissolved in the water. The mud at the bottom of the Puddle contains just the right amount of sulfate to furnish an essential element of our metabolism. There are periodic infusions of liquid into Little Puddle containing small amounts of dissolved carbon compounds which we use in constructing our bodies. Furthermore, the lower forms of life that inhabit our Puddle and form the basis of our food supply also depend on these same conditions. How could the beneficial methane-producing bacteria exist without the mud that shields them from the outer oxygen and gives them the raw materials for their metabolism? Where would the algae that live on the surface of Little Puddle go if the Puddle were much greater in extent, so that the surface was much farther from the mud-cushioned bottom? There may be other environments in the Universe, but it is impossible to imagine any form of life that could be adapted to the differences that exist between them and Little Puddle.

"We rotifers are so well adapted to Little Puddle," he continued, "that in any environment which did not possess all of the same qualities in exactly these proper amounts, rotiferian life could not exist. Since we are an essential part of the ecology of Little Puddle, if we could not survive in another environment, neither could the whole Puddlian biosphere.

"The conclusion is inescapable," he stated with finality. "Life is only possible in Little Puddle, or in other identical Puddles. The rest of the Universe is barren."

Another rotifer rose to second Philo and extend his argument.

"Since that is so," said the second, "it must be that the whole Universe is designed to ensure the existence of our home. Were there no depression in the surrounding rock, Little Puddle would not have formed, and we would not be here to appreciate it. If water were to freeze at a slightly higher temperature, Little Puddle would ice over and the nutrient rain that fructifies it could not get down to us. By using a new idea that I call the 'rotiferic principle,' I can demonstrate that the laws of nature must be exactly what they are and no different. Otherwise there would be no rotifers here to know about them. For example, if the heat necessary to vaporize water were slightly lower, then after the Big Rain created it, Little Puddle would have evaporated before the many hours that were needed for the first generation of rotifers to emerge from their eggs.

"So," he concluded, "but for the providentially high value of the heat of vaporization of water, the Universe would be empty of any life that could philosophize about it."

The listeners applauded this dazzling display of the power of rotifer reasoning, as did all the other inhabitants of Little Puddle, with the exception of a few insects on its surface who had seen some of their fellows eaten by a passing frog. Meanwhile, outside of Little Puddle, 10^{30} living beings followed their life-styles in a variety of Earth environments, oblivious to the rotifers' proofs of their nonexistence.

Chapter 5

The Origin of Earthlife:
An Unsolved Problem

Efforts to understand how life arose on Earth have been hindered by the absence of any direct evidence. The events of the past several billion years have erased all traces of very early life. We have only the end products of evolution to examine. It is as if we were trying to reconstruct the process used in ancient times to construct chariots by inspecting the workings of an assembly line for constructing automobiles. Given the lack of evidence, a number of speculative theories have been put forth. We will consider first a few such theories which we think are on the wrong track.

a. Spontaneous Generation

In past centuries many scientists believed that living organisms, even mammals, could be formed in simple ways from nonliving material. It was thought that decomposed organic matter, or even mud, would suffice to generate life. One typical seventeenth century recipe claimed to produce adult mice in twenty-one days from a mixture of wheat and sweaty underwear! As time passed, such extravagant claims were withdrawn or discarded. By the mid-nineteenth century, this theory of spontaneous generation was restricted to microorganisms. Further experiments by Louis Pasteur and others showed that even microscopic life arose only from preexisting life. This is not surprising. It is no more plausible to assume that a random group of molecules, even organic ones, would assemble to form a cell than it would be to expect a random group of human beings to collect spontaneously and assume the full workings of a stock exchange, or that a group of words shaken together randomly would form this book. But life has not existed forever, since the whole Universe is but ten to twenty billion years old. Somehow, at a stage in the past, there was a transition from a lifeless world to one in which there was life. Some state of matter must have existed that was simple enough to have been formed by random processes, yet complex enough that additional likely transformations would produce life. The formation of such a state was clearly a critical event.

CARBON COMPOUNDS FORM READILY BY NATURAL PROCESSES

While there is little we can say with certainty about the early composition of matter that gave rise to life, it is easy to guess that it involved compounds of carbon. Carbon compounds are so important in the present functioning of Earthlife that it is unlikely, though not impossible, that they were absent in the early stages. At one time it was thought that a critical problem concerning the origin of life was to explain how compounds of carbon could form in a lifeless environment. We now recognize that this process occurs easily in the Universe and poses no problem at all.

It is easy to see how this misunderstanding was created by historical circumstances. By the early nineteenth century, a large number of chemical substances had been isolated and purified from plant or animal sources. These compounds were usually liquids or solids with a low melting point. They were composed of carbon and hydrogen, sometimes with the addition of oxygen and/or nitrogen. When placed in contact with a flame, they burned readily. In all these properties they differed greatly from metals, salts, and other substances of mineral origin. The compounds derived from animal or plant sources were called organic, the others inorganic. For a time many chemists felt that organic compounds could be produced only in living cells, just as the cells themselves originate only from other cells. In 1828 a German chemist, Friedrich Wohler, produced urea, a component of urine, by heating ammonium cyanate, a substance believed to be inorganic. In a letter to Baron Jöns Jacob Berzelius, a strong follower of the theory that only living things could produce organic chemicals, Wohler wrote, "I must tell you that I can prepare urea without requiring a kidney or an animal, either man or dog." It was a great breakthrough at the time.

At present, it is accepted that all organic compounds can, in principle, be prepared in the laboratory from materials of nonliving origin, and many have been produced in this way. Furthermore, it has been found that the compounds found in living things are but a very few of the innumerable carbon compounds that have the same general properties. The term organic chemistry is reserved now for the study of these carbon-containing compounds, while biochemistry is concerned with the chemistry of living systems. The difference between life and nonlife does not reside in a fundamental difference in the chemicals of which life is made.

It is one thing, however, to recognize that organic substances can be prepared in the laboratory and yet another to realize that they can form spontaneously in nature. Within the past two decades, clear evidence has been obtained of the abundant presence of such compounds in at least two extraterrestrial locations: certain meteorites and interstellar dust clouds.

There is a class of meteorites called carbonaceous chondrites, containing several percent of carbon in compound form. The first investigations into the nature of these carbon compounds generated considerable excitement. Many biologically significant compounds were detected, including amino acids. The more biologically important of the two mirror-image forms of the amino acids was present in greater quantities. Later, more careful analyses revealed that the early results were due to biological contamination from improper handling of the samples on Earth. Subsequent studies were carried out with improved methods. The Murchison meteorite, which fell in Australia in 1969, received considerable attention. Analyses showed the existence of amino acids, but both mirror-image forms were present in equal amounts. A number of molecules important to Earthlife were present, but others were absent. The organic compounds contained in this meteorite, and others, constitute a very complex mixture, which, by every indication, has been formed by nonbiological processes; at least not by Earth biology.

A large part of the mass of our galaxy is composed of cold gas clouds that exist in the space between stars. It is possible to get some idea of the chemical composition of this gas by the use of radio telescopes. The technique is laborious. Organic molecules have several sets of "fingerprints" that can be used in their identification. One of these is the radiation that they emit. It is possible to measure or calculate the wavelengths of radiation emitted by a particular organic compound and then see whether some of these wavelengths are present in the signals coming from the interstellar gas. If so, it is strong evidence that the compound is present. By this method several dozen organic compounds have been detected in interstellar space. The largest one definitely identified so far has nine carbon atoms. More complicated molecules will surely be detected. It is still not certain how the compounds are formed in such a surprising environment. Their existence suggests that the formation of simple organic substances in nature must be a rather common event. It is easy then to presume that such events must have taken place on the primitive Earth.

THE PREBIOTIC SOUP

One of the dominant concepts in existing theories of life's origin is the "prebiotic soup." It is believed that the seas of the primitive Earth gradually accumulated organic material until they had a soup-like consistency. The organic molecules probably were originally produced in the atmosphere and rained into the oceans. Theories of the nature of the soup have varied according to the taste of the investigator: from chicken broth (hot and dilute) to vichyssoise (cold and concentrated). Leslie Orgel has estimated that the organic matter would be about one-third as concentrated as an average commercial chicken broth. The thought of an ocean of chicken soup may seem strange to us. In present times, it would quickly spoil and turn rancid. We must recall that neither molds nor oxygen were present on the early Earth to spoil the broth.

An important experiment supporting the existence of such a soup was first performed in 1953. Stanley Miller, a graduate student working with Nobel prizewinner Harold Urey, exposed a mixture of methane, hydrogen, ammonia, and water to an electric discharge. This experiment simulated the effect of a thunderstorm in the atmosphere of the primitive Earth. The complex mixture that formed was shown to contain a number of amino acids as well as many other organic compounds. Some of them are important in biology today, while others do not normally occur in living matter. Since that time, many other experiments of the same type (the term prebiotic is applied to them) have been carried out. It has been shown that compounds can be produced from mixtures of simple gases by a variety of the energy sources that may have been present on the primitive Earth: ultraviolet light (from Sun), heat (from radioactivity and volcanoes), shock waves (from thunderstorms and meteorites), and others. These studies further support the concept that a variety of organic compounds were present on the surface of the early Earth.

b. The Randomly Made Replicator:
A Miscalculation of the Odds

If we accept the picture described above, the main gap that remains to be closed in order to understand the origin of life, is the transition from the soup to the first replicator. A number of writers have assumed that this is a minor detail. If the soup were simply stirred long enough, the right circumstances would surely arise at random and the replicator emerge from the muck. One geology text has stated, "How many times 10,000 trials of each of many such random events could have occurred within a period of 3.3 billion years: . . . No one familiar with statistics rejects the idea of chance chemical combinations simply because there wasn't enough time. There was a huge abundance of it." Furthermore, with a little imagination we could speed things up in the lab today. It was stated in a *New York Times* article, "Whatever the case, many biologists believe it is only a matter of time before someone thinks up the right sequence of environmental conditions to which to subject an appropriate broth of non-living chemicals, creating life in the laboratory for the first time."

Alas, it is not that easy. The broth and the replicator are separated by the same sort of differences in order, as is a random collection of letters from the play *Hamlet*. In the evolution of life, there is no author to shape the letters into the play. Instead, it was the natural processes taking place on the early Earth, according to the laws of physics and chemistry, that made the soup into life. Much of the theory of how this occurred is unavailable, and a lot more knowledge is needed before we can cross the gap. Let us consider the difficulties to be overcome, in two possible scenarios. In one, the soup, once formed, is left unstirred. No further matter or energy is supplied to it, but the molecules continue to move about and interact with one another. In the second, a continual supply of energy stirs the soup.

If the sea of soup is left to itself, it would proceed toward the destination to which all isolated systems in the Universe tend to go: equilibrium. To understand equilibrium, let us think of it in human terms. Imagine a group of people, each with a different amount of money. Whenever two of the people meet, they compare their amounts, and the one with more gives half of the difference to the one with less. After everyone has had a few meetings, each person will have about the same sum of money. After that, further meetings will result in little or no transfer of money. This is what is called equilibrium.

Suppose now that the equilibrium is disturbed by one person finding a large sum of money. That person will have more money than the others for a while, but the transfer process will gradually restore equilibrium, with everyone having a greater amount of money. If there is a continuous flow of funds from the outside to some members of the group, at a rate similar to or faster than the rate at which it is transferred between them, then a situation could arise in which some people always have more money than others —a deviation from equilibrium. If we substitute the idea of energy for money and molecules for people, we can understand how a flow of energy can keep a group of molecules away from equilibrium.

Another analogy can be used to understand equilibrium in the situation where a group of different chemicals are mixed together and react with one another. Picture a dance hall, with the orchestra playing a popular tune. A large number of dancers are on the floor, others stand at the side or wait in line for refreshments. People are continually moving from one location to another, but the relative amount of people in each location remains constant. Overall they are at equilibrium.

We now change the conditions. The orchestra plays a much more difficult and rapid tune. For a time, there is confusion. Eventually, a new equilibrium is reached, with a smaller proportion of people on the dance floor. Equilibrium in this analogy involves an average distribution of the population between certain favored locations, while others (such as the ceiling!) are vacant. Similarly, equilibrium in a mixture of chemicals involves the presence of certain molecules, but not others.

For the prebiotic soup, the equilibrium state is mostly a mixture of those compounds from which it was created: methane, water, carbon dioxide, nitrogen, ammonia, hydrogen—all very simple molecules. At equilibrium, the amounts of amino acids and other molecules of interest to biology would be extremely small. They were formed in the prebiotic soup because external energy was applied, disturbing the equilibrium. This is analogous to the temporary inflow of money into our group of people. Left alone, the soup will gradually re-form the simple starting mixture or something very similar to it.

What chance is there that at equilibrium a simple molecule of replicator would form? (One might be enough.) Remember, we have oceans of material and billions of years of time. The answer is—almost none at all (about $10^{-1,000,000}$). The number is so small that it is of a type unfamiliar even to scientists, who deal with very large and very small numbers. Let us think of it as a bet. We will improve the odds by putting every atom of matter in the Universe into the soup, permitting the atoms to interact with each other as rapidly as possible within the laws of physics, and allowing all the time since the creation of the Universe for the replicator (which we will assume to be a nucleic acid of moderate size) to form. You bet one cent that at least one replicator would arise at random in the equilibrium mix. If you won, how large would the payoff have to be in order to make the bet fair? The answer is that it could not be done. Even if there were enough gold to fill all space out to the rim of the Universe, this amount would not be nearly enough to justly reward the person who made the one cent bet that a replicating molecule would form by chance. The same conclusion would apply to the random formation of any other specific very large, complex organic molecule.

Even the lifetime of the Universe is nowhere near sufficient for a replicating molecule to form by chance under equilibrium conditions.

The second scenario is much more likely—that life occurred under conditions that are very far from equilibrium. Fortunately, the Sun has been a very reliable source of energy. Ultimately, the energy emitted by the Sun is dissipated into the void of outer space. While en route, a portion of it passes through the atmosphere and reaches the sea and land surfaces of the Earth. As shown in prebiotic experiments, this energy can interact with the primitive atmosphere to form the prebiotic soup. It can further cause the formation of more complex molecules from the simpler ones initially present in the soup. However, though the formation of larger molecules is quite possible in this situation, the random formation of a nucleic acid or protein replicator is still extremely unlikely. The reason is that too many types of complex organic molecules can exist. To understand this, we must appreciate the immense number of possibilities that are inherent in organic structures.

Let us recall the hydrocarbons, the simple class of compounds which are made only of carbon and hydrogen and comprise gasoline. One way of classifying them is according to the number of carbon atoms they contain. Only one hydrocarbon exists that has one carbon atom, methane (CH_4), which we met earlier. There are three different hydrocarbons that have two carbon atoms, with formulas C_2H_2, C_2H_4, and C_2H_6. As the number of carbon atoms increases, the number of possibilities goes up rapidly. The introduction of other elements also serves to increase possibilities. For example, only one hydrocarbon with one carbon atom exists, but we know of half a dozen compounds that have oxygen in addition to hydrogen and one carbon. While in some cases there are large differences in the chemical energy of the compounds, and so in the likelihood of their production, there are other cases in which very different compounds could be produced with equal ease.

The situation is complicated further by the existence of isomerism. This term describes the situation that occurs when more than one compound contains the same number of each type of atom in it. For example, two different compounds, ethyl alcohol and dimethyl ether, are both described by the formula C_2H_6O. Ethyl alcohol is the familiar intoxicating component of wines and spirits, while dimethyl ether is a nonintoxicating gas used in refrigeration. The two differ in the arrangement of their atoms as well as in other properties that depend on this arrangement (Fig. 13). Such compounds are called isomers.

$$
\begin{array}{cc}
\overset{\displaystyle H}{\underset{\displaystyle H}{H-C}}-\overset{\displaystyle H}{\underset{\displaystyle H}{C}}-O-H \qquad\qquad H-\overset{\displaystyle H}{\underset{\displaystyle H}{C}}-O-\overset{\displaystyle H}{\underset{\displaystyle H}{C}}-H
\end{array}
$$

ETHYL ALCOHOL   METHYL ETHER

Figure 13. The arrangement of atoms in two isomers

The number of possible isomers of an organic formula goes up rapidly as the number of carbon (and nitrogen and oxygen) atoms increases. Thus, the formula $C_4H_8O_2$ has over a hundred isomers, and $C_4H_8N_2O_2$ has well over a thousand. The situation is analogous to planning the seating arrangements for guests at a dinner table. If you have one guest, only one arrangement exists. With two guests and yourself, two arrangements are possible, which are mirror images of each other. The two are about the same for conversational purposes. However, if one guest tends to kick with his right foot and you have a sore left shin, you will notice the difference between the two seating plans. If four people are seated at dinner, six arrangements will result. By the time your proposed dinner has grown to include nine guests and yourself, over 300,000 possible seating plans will exist.

While this number is large, it is small compared with the one you will have if you are uncertain of the exact identity of your guests. If in your dinner for ten, you must select nine guests from a possible list of twelve and wish to consider the various seating arrangements, there will be more than eighty million to choose from! Similarly, in organic chemistry if you considered all compounds with four carbons and any number of H, N and O atoms, the possibilities would extend to many tens, or perhaps hundreds, of thousands.

Let us now return to the prebiotic soup. Suppose that a benevolent flow of energy has allowed the random synthesis of various small organic molecules up to the size of nucleic-acid building blocks. The smallest such building block, a nucleotide, has the formula $C_9H_{14}N_3O_7P$. The number of organic isomers that can be constructed using those elements and up to nine carbon atoms will be very large, at least in the tens of millions. Many of these compounds will not be stable; however, our nucleic acid component will still be one of millions of such chemicals in the soup. But a component of a nucleic acid is no closer to being a replicator than a fragment of glass is to being a wine decanter. In order to function, the pieces must be put together in a particular way. For nucleic acid components, two units must be joined to form a larger unit, plus a molecule of water. This process is sometimes called "splitting out water." It can be achieved by applying energy such as heat. Nucleotides can join by splitting out water in a number of ways, but only one leads to the structure known to be involved in Earthlife. Unfortunately, many of the small molecules present in the prebiotic soup can also connect to one another, and to nucleotides, by splitting out water. The total number of possibilities involved is enormous.

In this situation what are the odds against even a few dozen nucleotides avoiding other reactions, finding each other in the mixture, and joining properly to form a mini-replicator? The fair payoff on a wager of such an event would have to resemble the one already described for an equilibrium state: The odds against the formation of a replicator resembling any of the nucleic acids or protein that we know today are extremely large. What would happen is that various complex molecules would be formed, containing chains, branched chains, and rings. Each such molecule would be made from a mixture of the small molecules present in the soup; for the most part these are neither nucleotides nor amino acids. It is possible that a very small chain of nucleotides, perhaps ten units or so in length, might be formed by random combinations in this mixture as a rare event. It is conceivable also that this small chain might accidentally have the power to perpetuate itself in the midst of molecular chaos, although we have seen that the smallest known replicating units are much larger. If this scenario did occur, then it may be possible to simulate it in well-devised laboratory experiments. However, the chances of it being true do not seem great. With some confidence, we can turn away from the possibility of a randomly formed replicator and consider other ideas.

c. The Predestinists

The advocates of the random replicator can be regarded as guardians of the last bastion of spontaneous generation. As we have discussed, the weakness of their position can be demonstrated on statistical grounds. In addition, the random nature of the mechanism has little appeal to many others, who would prefer to believe in a more purposeful Universe. According to this school, the events leading to the present state of life on Earth were inevitable, given the laws of the Universe. A built-in bias guaranteed that the final result would be just as it is. In the broadest view, not only the biochemical origin of life on Earth but even the values of the fundamental physical constants of nature were preselected.

We would like to use the term "predestinist" in this book to refer to those who believe that all or a portion of the events leading to the present state of terrestrial life were inevitable, given the laws of the Universe. In some cases, exponents of this viewpoint treat the course of events as though it were part of a preexisting plan with theological implications. We will use the term for those who believe in the existence of a prior plan and for those who do not. We are not concerned with theology here, but with science. We cannot demonstrate that all such concepts are wrong. Some speculations, such as those involving alterations of the fundamental physical constants, cannot even be tested. In other cases, however, predestinist expectations can be compared with experimental fact. When this is done, they are generally found wanting. Unfortunately, many predestinists seem to be unconscious of their views. Their assumptions are taken for granted, then used in the interpretation of experimental facts and the design of additional experiments. The consequence is that certain predestinist myths have been accepted as established facts by other scientists and the public in general. The result is a very limited view of the possible forms that life may take on a planet like Earth and in the Universe as a whole.

At the biochemical level, predestinists assume that the synthesis of organic molecules in the prebiotic soup did not take place at random, but led preferentially to the formation of the small organic molecules essential to Earthlife. An experiment crucial to this point of view was the one performed by Miller and Urey described earlier. This experiment demonstrated that amino acids and other small organic molecules were formed when a mixture of gases that may represent the composition of the primitive Earth atmosphere was exposed to an energy source. The very appearance of amino acids seems, to devoted predestinists, a conclusive demonstration of the validity of their faith. For those with less conviction, it is important to look at the situation in more detail.

For a clue as to what actually did occur, we are indebted to Stanley Miller himself. In a published account of the circumstances surrounding his famous experiment, he reports an unsuccessful first attempt, "The next morning there was a thin layer of hydrocarbon on the surface of the water, and after several days the hydrocarbon layer was somewhat thicker." No amino acids at all could be detected upon analysis of the mixture. Miller then interchanged the location of the condenser and spark in his apparatus, and repeated the experiment (the previous arrangement of the apparatus had been selected with safety considerations in mind). Good quantities of amino acids were produced. Writing twenty years later, Miller could state, "It was also surprising that the yields of amino acids from the first experiments are the highest so far reported in any prebiotic experiment of this type."

It is a remarkable *tour de force* to achieve the best and worst possible results in the first two attempts at an experiment, but this does not convince us of the inevitability of amino acid formation. There is unfortunately a dearth of information in the literature about the types of primitive chemical mixtures that do not form amino acids upon exposure to a source of energy. There has also been no systematic survey of all the other compounds which are without biological importance that are formed. This is not the type of result that makes a subject for publication. It has been learned in these experiments that amino acids are not formed if conditions in the mixture are oxidizing. This is considered irrelevant to the origin of life on Earth, as a reducing atmosphere was present at that time. However, ideas about the exact composition of Earth's early atmosphere have changed with time, and many of the early experiments reporting the formation of amino acids may also be irrelevant, because they begin with compounds not present in that atmosphere.

The predestinist point of view, while not impossible, is simply not supported by any systematic body of experimental fact. In order to establish that there is an inevitable trend in nature toward the formation of biologically important amino acids, a series of experiments would have to be run, testing the effects of all the variables such as temperature, concentration, geological environment, etc. The yields of products that are both important and irrelevant to biology would have to be compared thoroughly in these cases. This has certainly not been done.

If we examine with care the positive results that have been reported, some general rules emerge. The amino acids that are formed are the smaller, simpler ones. Small amino acids, both those important to Earthlife and those absent from Earthlife, are generated. The most complex amino acids prominent in biology are not observed. In general, one would expect small, stable organic compounds of all types to be formed in higher yields as a product of these experiments, and more complex ones in lower yields, regardless of their relevance to biology. When one has a large number of variables at one's disposal, they can always be adjusted empirically to favor the formation of a particular class of compounds of interest, such as amino acids.

Perhaps another analogy will be helpful. Imagine again that many letters are jumbled together and are blindly selected from the pile in groups of five. Once in a while, an English word such as SMILE will be selected. Usually, a word in another language such as BONNE or a nonsense word such as GPNIZ will be chosen. Obviously, in order to establish a thesis that some selection procedure tended to produce English words, it is not sufficient just to cite the few English words chosen. It would be necessary to give the complete array of words selected in a large number of tries and analyze them to determine whether English words occur more often than chance would suggest. It is just this analysis that has not been done with prebiotic synthesis.

The smallest complete unit of a nucleic acid is a nucleotide. No nucleotide has yet been produced by exposing a mixture of simple atmospheric components and salts to a source of energy. This is understandable, as nucleotides are much more complex than the simplest amino acids.

Despite the massive shortcomings in the demonstration of the prebiotic synthesis of the small molecules important to

Earthlife, others have gone on to demonstrate how the monomers could combine to form polymers. Abandoning the complex mixtures produced by the initial experiments with simulated atmospheres, the experimenters have worked with purified amino acids or nucleotides. They have exposed the monomers to a variety of simulated prebiotic conditions and succeeded in preparing, in some cases, small proteins or nucleic acids. In order to justify the conditions under which these experiments were performed, various imaginative scenarios for the primitive Earth have been invoked: freezing oceans, boiling volcanoes, changeable hot and cold deserts, evaporating ponds, and others.

These experiments have obvious flaws. First, no evidence is presented to support the existence on the primitive Earth of the conditions that were selected rather than any others; these conditions are used because they yield the desired results. An even more serious shortcoming is the use of pure starting materials, rather than mixtures simulating the prebiotic soup. The production of proteins or nucleic acids from pure amino acids or nucleotides is about as surprising as drawing a flush poker hand from a deck which contains only one suit. This result has little relevance to the situation in nature, where a complex mixture of many substances was probably present. In particular, it seriously misleads us about the difficulty in generating nucleotide polymers in the conditions of early Earth. What it does show is that the monomers can form polymers without the need for biological catalysts. This is important to know, since otherwise it is hard to see how many macromolecules could ever form prebiotically.

In summary, simulation experiments in the laboratory do little to suggest the inevitability of the formation of the molecules important to Earthlife. They demonstrate that certain of these molecules can be formed under specified conditions. When the conditions vary, other types of product may predominate. Even in the best cases, a complex mixture is the result. Even if amino acids and nucleotides were present in this mixture, they would still have great difficulty in finding one another and combining in the proper manner. If the predestinist scenario is to be plausible, some inevitable process must be demonstrated that brings them together and connects them properly. This has not been done. As we shall soon see, however, the problem can be approached from another direction—if we allow the possibility of evolution in the prebiotic soup before individual living things arose.

COSMIC PREDESTINISTS

Despite the weakness in the basic experimental foundations of predestinist thought, others have used these foundations to erect towers that extend to the end of the Universe. Let us quote, for example, Professor Norman Horowitz, a biologist at the California Institute of Technology at Pasadena:

> *The implication of these results is that organic syntheses in the universe have a direction that favors the production of amino acids, purines, pyrimidines and sugars: the building blocks of proteins and nucleic acids. Taken in conjunction with the cosmic abundances of the light elements, this suggests that life everywhere will be based not only on carbon chemistry, but on carbon chemistry similar to (although not necessarily identical with) our own.*

The allusion to cosmic abundances refers to the fact that the three most abundant elements in Earthlife—carbon, hydrogen, and oxygen—are also three of the most prevalent in the Universe. It is interesting that carbon is not nearly as relatively abundant on the surface of the Earth as in the Universe. On the other hand, iron, silicon, and aluminum (very common on Earth's surface) are present in very low percentages in Earthlife.

The support for the extension of predestinist ideas from the prebiotic Earth to outer space comes from the discovery of organic compounds in carbonaceous chondrite meteorites and interstellar dust clouds, as we described above. The meteorites are said by predestinists to be rich in the molecules of life, while the interstellar molecules are depicted as the precursors of biomolecules. Some scientific writers have drawn the further conclusion that Earthlife, or at least the key molecules characteristic of it, may have originated in outer space and then been transmitted to this planet, not in the distant past, as we describe below, but even today.

Leading exponents of one such view are the British astronomer Professor Fred Hoyle, together with his colleague, Dr. C. Wickramsinghe. They claim that "the essential building blocks of life—amino acids, nitrogen-bearing heterocycle compounds and polysaccharides—are formed in space. These compounds occur in large quantities throughout the galaxy." In particular, they postulate that the processes leading to the origin of Earthlife take place repeatedly within comets, culminating in the formation of functioning bacteria and viruses. They claim that a collision of a comet with Earth may have initiated life here, while subsequent collisions introduced epidemics of influenza and other diseases.

These concepts are imaginative but unfortunately not supported by the facts. Meteorite analyses have revealed a complex mixture of compounds, with no demonstrated preference for Earth biochemistry. The organic molecules detected in space thus far are much smaller and simpler than those produced in the Miller-Urey type of experiments. Up to 1979 no amino acids have been reported. The molecules detected make up a rather unearthly collection, as might be expected. They tend to involve long, hydrogen-deficient chains of carbon atoms, a type not found in the biochemistry of Earthlife. Many such molecules which would be very short-lived on Earth are quite stable in the vacuum between the stars. Die-hard predestinists may choose to view this collection as the precursors of biological molecules. This is true only in the sense that they may be the precursors of anything.

It is a textbook exercise in college organic chemistry to devise suitable pathways for the conversion of small organic molecules into almost any desired product.

What about the possibility of epidemics from outer space? The notion that viruses evolve in comets and then can infect organisms on Earth poses many serious problems. One is that the viruses themselves could not function as living things in the comets, since they require cells to replicate. There would be no evolutionary pressure for certain viruses to develop the mechanisms for entering a cell and capturing the cell's genetic machinery, since the viruses in the comet would not be influenced by what happens to those that reach Earth. None of the mechanisms for evolution that are available to viruses on Earth can exist in comets, and it is very hard to imagine any reason why such evolution would occur there. Since the influenza epidemics that Hoyle and Wickramsinghe are trying to explain are the result of a viral evolution that avoids the antibody resistance to previous strains of influenza, their explanation seems untenable. Its likelihood ranks with the probability that Christopher Columbus, in his landing on San Salvador, would have found that Spanish was the indigenous language of the natives.

These efforts to picture the Universe as an incubator for the biological molecules of Earthlife still do not represent the fullest scope of predestinist thought. There are others who would extend the predestined course of events back further to the origin of the Universe. A Boston University physicist has written:

> *An examination of the fundamental laws of physics and the values of the constants associated with these laws reveals an astonishing predisposition of nature for the appearance of life . . . even small variations of the physical constants would seriously hamper the origin and subsequent evolution of life anywhere in the cosmos. As a result, one is almost led to conclude that our*

universe was predestined, from its very inception, to harbor intelligent life, which consequently must be of special cosmic importance.

This is an example of a view that has been advanced in recent years by several astrophysicists under the name of "the anthropic principle." Paul Davies states it as follows:

Some scientists have become intrigued with the idea that our own existence may not be a random event in the universe but one dependent on the presence of very special circumstances. This implies that some of the features of the universe which we observe cannot be separated from the fact that we are alive to observe them, that life is very delicately balanced in the scales of chance. Our very existence severely restricts the nature of the universe, which must fit into what appears to be a very small class, the class of universes in which we can conceivably survive.

We think that anthropic arguments are seriously flawed in several respects. For example, they contain the assumption that only Earthlife exists and has developed intelligence, a view unsupported by any real evidence. If, as we shall argue, many other types of life exist in the Universe, a change in physics that made Earthlife impossible would not necessarily eliminate all types of intelligence from the Universe. Indeed, such a change might even allow some new types of intelligence to occur that cannot exist in the present Universe. In this respect, users of the anthropic principle suffer from a severe lack of imagination in working out the consequences of alternative laws of physics.

An even more serious flaw is in the underlying logic of the anthropic principle. Not only does the existence of intelligent Earthlife depend on the laws of nature being exactly what they are, but so does almost everything else about the Universe. We might as well start with any other fact, such as the existence of a particular vein of copper ore, three hundred meters beneath the ground in Greenland, and use a "cupric principle" to infer that the laws of the Universe are designed to ensure the existence of that vein.

Another analogy may help to clarify the problem in this type of reasoning. If the Battle of Waterloo had had a different outcome, it is unlikely that any reader of this book would exist, as the particular combination of events that led to each of us being here is so unlikely that any large change in what happened in 1815 would almost certainly have led to a different group of people being alive today. But it would be absurd for reader John Jones of 12 Cedar Lane, Topeka, Kansas, to infer just from his existence that the Battle of Waterloo took place and was lost by Napoleon. It would be equally absurd to try to infer from the Battle of Waterloo that Mr. Jones would be reading this book 165 years later. Only after we know both of these facts are true can we construct a link between them.

Similarly, just by knowing that intelligent Earthlife exists, we are no more able to infer what the laws of nature are than from knowing any other individual facts. Even less can we take the known laws of nature and deduce that intelligent life would develop on Earth after four and a half billion years. The only way we could learn something from the anthropic principle is if we could show that among all possible laws and all values of the constants of physics, only one set can lead to intelligent life, the set we have now. As we have indicated, there is no reason to think that this is so. Instead of being a way of discovering the laws of nature, the anthropic principle is only a benediction said over these laws after they have been discovered by other means.

d. An Extraterrestrial Origin for Earthlife?

We have seen the difficulties that beset many popular theories concerning the origin of life on Earth. Because of these problems, a different and fascinating solution to the puzzle has been suggested. Our type of life may be unique on Earth because it arrived from elsewhere. It may have originated on another world among a host of competitors, but been the sole survivor of the process which transported it here. This hypothesis, called "panspermia," was first suggested by the Swedish chemist Svante Arrhenius in 1907. It is incomplete, however, without an account of how these first cells arrived here. Originally, it was suggested that spores may have drifted out of the upper atmosphere of an inhabited planet, and been transported through interstellar space by the pressure of light. This suggestion has been criticized by Carl Sagan, who demonstrated that such spores, if at all like those we know on Earth today, would be killed by cosmic radiation. Spores could, of course, be protected from radiation by being encased in a meteorite. Again, Sagan has demonstrated that it is unlikely that even a single meteorite from another solar system has reached the Earth during its entire history. The chance that such a meteorite would carry viable life forms is much less.

If one rules out the accidental transmission of extraterrestrial life to Earth, the possibility of purposeful "infection" of Earth remains. Francis Crick and Leslie Orgel have coined the term "directed panspermia" for this scenario. They suggest that an intelligent civilization may have been motivated to spread life throughout the Universe and would have launched specialized spacecraft for that purpose. The craft would have contained bacterial spores, shielded from radiation and otherwise prepared to endure a voyage of thousands or millions of years. When a suitable planet was found, the spores would be delivered to the surface.

The variations on this theme that are possible are also striking, if sometimes less flattering. For example, Thomas Gold, a Cornell astronomer, suggested that the first living organisms may have been planted here inadvertently, perhaps in the garbage of cosmic visitors.

The difficulty with these proposals is that they do not solve the problem of the origin of life, but merely remove it to a distant locale. There is no indication that another environment would be more suitable for the formation of Earthlife than the early Earth itself. Furthermore, the extra time gained if life originated elsewhere is not great enough to permit a purely random origin for it. At present, there is no reason to believe that our life could have originated more easily elsewhere than here. Nor is there any evidence that viable spores are present in outer space. Without such evidence, theories that explain how life originated locally must be given priority.

e. Gradual Chemical Evolution

After considering a number of ideas about the origin of Earthlife, we find ourselves back where we started. Yet we are still confronted with one obvious fact: Life exists here and presumably started here—somehow. How can this have happened without a near-miracle? An alternative possibility remains: It arose as the result of a gradual evolutionary process. Just as human beings have come to exist not by a sudden random association of cells but by natural selection working through stages of increasing complexity, so the first replicator did not arise from a mixture of molecules by a random event but as the end product of a number of stages of increasing complexity. The creation of the replicator was an important step in evolution; however, it was not the first one.

What processes preceded the evolution of the replicator? Not Darwinian natural selection, since advantageous changes were not propagated by reproduction. However, some manner of chemical self-organization could have occurred under the prevalent conditions. The chemical compounds present in the prebiotic soup interacted randomly with one another, inducing a wide variety of chemical reactions. Under the environmental conditions existing then, some of these reactions produced new compounds that enhanced the concentration of molecules in the soup. Other synthesized compounds had the opposite effects. There would therefore be a tendency for these primitive catalysts and their components to occur to a greater extent than if they were formed only by random processes. Under these conditions, the formation of macromolecules made of a few types of simple units would more easily come about.

Although this process of chemical evolution leads to the same results as the predestinist scenario, there is an important difference between them. The outcome of the chemical evolutionary process is much less determined than the predestinists would want. Small changes in the initial concentration of chemicals or in the environment during this type of evolution could lead to quite different end results, in the sense of biochemistries that are quite different from those of Earthlife. In the predestinist scenario, it is Earthlife or no life that must emerge.

An analogy using letter combinations can illustrate how chemical evolution might occur. Imagine a prebiotic alphabet soup, containing letters rather than chemicals, and circumstances which favor the joining of letters to form pairs and the pairs to form four-letter words. If this were done at random, the word MAMA would be formed once in every 457,000 four-letter words. Suppose, though, that the pair MA had a catalytic effect. It speeded up greatly the joining of a consonant to a vowel, but not a consonant to a consonant, a vowel to a vowel, or a vowel preceding a consonant. One would get more MA's and PA's and LU's, and less AM's and ZX's. If the catalytic power of the MA's were having full effect, the fraction of MAMA's made would rise to one in every eleven thousand four-letter words. The COME's and the HERE's and the PAPA's would benefit to an equal extent. Let us now suppose that PAPA had an accidental catalytic power. It greatly favored the union of identical pairs to form four-letter words. The words MAMA and PAPA would now comprise almost 1 percent each of the new production. Several more stages of this type could lead to a single message and its constituent words becoming far more prevalent than any of its possible competitors. This message would then perpetuate itself by its own catalytic action on the words. The replicator has arrived on the scene.

Two requirements for such a self-organizing system are that the system be far from equilibrium and be subjected to a flow of energy. Other requirements probably exist, but they have not yet been explored. Scientists have found some examples of nonliving systems that show parallels to this self-organization and evolve in the direction of increasing complexity. One example is a chemical system that undergoes reactions leading to colorful rings and spirals which appear and disappear over and over (see page 140). Another is the emission of light by a laser. Efforts to describe such systems and their living analogues have been made mathematically by a number of scientists, including Ilya Prigogine, Manfred Eigen, and Hermann Haken. The mathematics used is found to be somewhat similar for the different systems, suggesting that some general principles may be involved. However, this subject is still in its infancy and is not yet capable of suggesting new systems in which similar behavior might occur. Therefore, it is not yet useful for predicting what environments might favor self-organizing behavior that could lead to the development of life.

Our empirical knowledge in this area is also very incomplete. Little has been done to develop self-organizing systems in the laboratory. Above all, we do not know what the minimum requirements are for the initial state, which sets the rest of the chain underway. Neither do we know how many different systems of this type are possible, nor the probability that any of them will develop into actual life. We will return to these questions again in later chapters. For the present, we find that our present picture of the history of Earthlife, while fascinating in itself, cannot decide how life may develop in other environments different from Earth, or how long such development might take.

Our analysis has not been entirely negative, however. We have cleared away certain prevalent, but flawed, ideas such as the predestinist theories, so that our minds are free to consider additional ideas for life on Earthlike planets and on Earth itself.

There is also an important implication to be drawn from the idea of pre-Darwinian chemical evolution, which is that the Earth's biosphere existed before individual living things. The chemical reactions that actually led to the first living things were similar to those occurring in living organisms today, although not as complex. The primitive cycles, driven by sunlight, were enough to define a simple biosphere of many interdependent parts. Out of this primitive biosphere the first individual living creatures arose.

If the Earth's biosphere did exist before individual living things, then we must consider that the same possibility can occur elsewhere. This suggests that our thinking about extra-terrestrial life should focus on alternative biospheres as well as on alternative living creatures. Furthermore, the very definition of life must take into account the idea that the biosphere was here first and that individual life forms are an outgrowth of it. It is time that we turn from the study of Earthlife to the more general questions of the definition of life and the basic conditions for it to emerge elsewhere.

LIFE

For reasons of professional pride . . . many biologists like to believe that there is an absolute categorical distinction between the biological and physical sciences, but their beliefs have not been gratified.
—P. B. MEDAWAR AND J. S. MEDAWAR

Spirals that appear spontaneously in certain chemical reactions indicate how order can arise in a nonbiological system."

Chapter 6

What Is Life?

From a "commonsense" viewpoint, nothing seems easier than to tell what is alive and what is not. Humans, birds, plants, and even bacteria look and act very differently from rocks, sand, and other nonliving things on Earth. The difference was underlined in a very funny way a few years ago when Pet Rocks were offered for sale in a number of stores. Pet Rocks were excellent at obeying certain familiar dog commands, such as "Sit" or "Stay." Getting them to "Come here" was naturally a bit harder to do.

While it is easy to describe the difference in behavior between a dog and a rock, it is surprisingly hard to capture the difference between the two in a precise definition. Especially when we try to extend our concepts and observations to less-familiar examples of Earthlife, we run into difficulties. A recent magazine article described the ocean floor as viewed by time-lapse photography:

> *Something that for three months had looked like a rock got up and moved about a foot, then settled down again and looked like a rock for three more months. Another rocklike thing sprouted an arm and waved it about for 12 hours, then remained motionless for the rest of the six months.*

For another example that is difficult to define, let us consider the tardigrade, a barely visible insect-like animal. Tardigrades can be dried out and stored in a bottle on a shelf for many years, to all intents like the lifeless ashes of Great-uncle Lemuel. But add water to instant tardigrade, and the animals will be restored to their normal state and scamper merrily about, no worse for their long, dry nap. Is tardigrade powder alive? It is an example of what has been called cryptobiosis, hidden life. If the powder is not alive, we have spontaneous generation just by adding a few drops of water, a view that biologists would consider absurd. If it is alive, then continuity of metabolism is not essential in order for us to consider something to be alive.

A similar problem exists with many viruses. They can remain in a powdery state on the shelf for years. Unlike tardigrades, they are not restored to full activity by the addition of water, but require the presence of host cells. When given that essential, the viruses carry out life functions and reproduce merrily, giving rise to many future generations. Are viruses alive?

THE NEED FOR A DEFINITION OF LIFE

In the face of such problems, there have been many debates among scientists and philosophers on the definition of life. No clear consensus on the answer has emerged. Some biologists, partly in exasperation, have declared the question to be meaningless, or worse. We do not wish to stir the ashes of past debates but rather to light our own new fire to illuminate that corner of the subject which is important to our topic: extraterrestrial life. It is an element of scientific faith that there are essential aspects of Earthlife which also apply to what we find elsewhere in the Universe. Depending on which aspects of life turn out to be the essential ones, prospects for extraterrestrial life will wax or wane. It is these aspects that our definition must capture. Such a definition would help us to recognize other systems, which while composed of very different materials than Earthlife, operate according to similar scientific principles. If we find such systems and if their overall behavior captures our interest, we would probably want to consider them alive.

Our definition of life must begin with what we know of Earthlife, and we must sort out those things that we recognize as being alive from those that we do not. But this definition will have to distinguish clearly what is essential to life from what is an incidental result of the history of life on Earth.

For example, some authors, calling upon the discoveries of several hundred years of biology, give limited structural definitions of life, such as, "A living thing is a cell or a group of cells, containing nucleic acids and proteins." This definition has the merit of accurately describing some properties of all the free living things that we have examined carefully thus far. But, if taken literally, it also implies that nucleic acids and proteins are essential to life and that no other basis of organization for life is possible. If this definition is extended uncritically to life beyond Earth, we should not expect to find it where those molecules are unstable. Since almost all of the Universe consists of such environments, this means we would have to ignore most of the Universe as possible homes for life. Suppose, however, that things can exist elsewhere with most of the other attributes associated with life, but which are not made of cells containing nucleic acid and proteins. The above definition would then be seriously misleading. It could cause us to dismiss systems that have lifelike properties and to disregard environments where these systems occur because no such life could satisfy the definition.

We think that this point is crucial, so we will emphasize it with another analogy. Imagine a city in which all taxicabs are painted yellow and no other yellow cars are allowed. A definition of taxicab for that city could then be given as: A taxicab is a yellow car. This definition would indeed single out taxicabs from all other cars in the city, but it would certainly miss the essence of "taxiness." It would also lead to great confusion if you traveled to another city where taxicabs were by law painted black. While it was certainly an important discovery in biology that known Earthlife is composed of cells containing nucleic acid and proteins, further analysis is required to decide whether this is a necessary aspect of all life, even on Earth.

It is possible that this analysis will show that under conditions at all similar to those on Earth, nucleic acids and proteins are either a unique or almost unique basis for life. In that case the identification of life with nucleic acids and proteins will be much more fundamental than we now know it to be, or somewhat more analogous to identifying the quality of "taxiness" with vehicles that transport small numbers of passengers for a fee. We think it more likely that nucleic acids and proteins are to life somewhat as a four-wheeled vehicle with a meter and an internal combustion engine is to "taxiness." That is, nucleic acids and proteins are the basis of the overwhelming majority of life forms in our time and place, but in very different times and places such life forms may be as uncommon as self-propelled taxis were in Hong Kong in 1850, and the basis for life there may be as different from nucleic acid and proteins as a rickshaw is from an automobile.

While we do not want our general definition to be so narrow and Earth-centered as to define away most extraterrestrial life, we also do not want it to be so broad as to be meaningless. It would serve little purpose to declare that life is matter that is composed of protons, neutrons, and electrons. This view would render the entire Universe alive and would be as useless as one that limited life to Earth. Life is the most interesting form of the organization of matter on Earth. Elsewhere in the Universe, or even on Earth, there may exist interesting ongoing processes involving matter and energy which capture our attention but are sufficiently different in how they are made and in what they do that we would not wish to apply the term life to them. The definition that we seek is one which, applied to the various objects that we can observe in the Universe, will sort them into the living and nonliving in a meaningful way.

A good definition of life has another value in aiding the search for extraterrestrial life. It can help suggest what we should look for. Even in our experience with Earthlife, prominent organisms that have not been anticipated have often been missed entirely. Two microbiologists, when questioned why a particular species had been overlooked for a long time, replied, "There was no a priori reason to expect the existence of this unusual creature."

Such situations are even more likely to occur in extraterrestrial exploration. Any searches that we carry out in the near future will be with automated probes having a limited capability for gathering data and for processing that data. The success or failure of such probes will almost certainly depend on giving careful advance consideration to what we are seeking. Whereas human beings in alien environments might be able to recognize exotic life forms by some intuitive synthesis of data, no such feat can be expected from equipment designed to gather a limited amount of data. The ambiguities in interpreting the results of the Viking Mars probe will give ample evidence of the problems of deciding when you have found something if you don't know what it is you are looking for.

To most of us, it is what living things do that is the interesting feature about them and what distinguishes them from the nonliving. For this reason, people are apt to prefer a definition of life that emphasizes function, and indeed gross function, of a type which is immediately apparent. One of the standard definitions of life is along these lines, identifying a living organism as something capable of spontaneous movement, growth, response to stimulation, and reproduction. Such a definition is hard to maintain when one goes beyond the large animals familiar to us, and is unlikely to be of much value in dealing with extraterrestrial life. It is not difficult to think of some organisms which fail to display one or more of these functions and yet would usually be classified as living, as well as some that satisfy all the functions but would not be considered alive. Thus, a spayed dog, or a mule, are no less alive for being incapable of reproduction. Plants usually do not move spontaneously, although they do grow. Nerve cells do not reproduce after a certain stage in their development.

On the other hand, flames display all the functions mentioned. They can grow happily when offered fuel, respond dramatically to wind and rain, and, when sparks are carried by the wind to a suitable new location, they can reproduce. A salt crystal growing in a solution also satisfies many of the criteria of this definition. It is hard to avoid the conclusion that while these functions are often indicators of life, they are not its essence, which must be sought elsewhere.

Another standard definition of life emphasizes its genetic and evolutionary aspects. According to this definition, life is something that reproduces, mutates, and reproduces its mutations. This does describe some aspects of the living things we know. But it is not obvious that the definition would adequately describe living things elsewhere. Faithful reproduction of mutations is a tactic that Earthlife has used to adapt to environmental change. Even on Earth it is not the only such tactic, since many living things change their environment rather than mutate. Also, in an environment that is static, mutation could be an irrelevant tactic. Furthermore, it is not clear if the faithful reproduction of random mutations is the most effective way to produce an altered living thing that is well adapted to a new environment. It might be simpler for a living organism to alter itself directly when confronted with an environmental change, rather than for it to produce altered descendants. This direct response could be a more effective one, even if it were tied to a less-faithful form of reproduction.

OUR DEFINITION OF LIFE

It is time to present our own solution. It comes in two parts. In one part we deal with the need to define the proper unit for recognizing life. We believe that it is the entire biosphere that best suits this purpose. The second part concerns the vital activity of life that distinguishes it from nonlife. We argue that this is the preservation and gradual increase of life's internal order. The next sections will deal with these questions in detail. At this point we will simply state the definition.

Life is fundamentally the activity of a biosphere. A biosphere is a highly ordered system of matter and energy characterized by complex cycles that maintain or gradually increase the order of the system through an exchange of energy with its environment.

THE BIOSPHERE AS THE UNIT OF LIFE

Many attempted definitions of life have run into difficulties because they failed to distinguish the part from the whole, and to decide at which level the term life should apply. Are we alive? Surely. But then, what about our organs or our individual cells? They can and do function as living objects for a time, even when separated from the remainder of our bodies. Viruses, and some subcellular components such as mitochondria, also perform many of the functions associated with life. We might also want to think of them as living, even though they can only function in symbiosis with the rest of the cell. Going further, certain small pieces of DNA can be absorbed by bacterial cells and transform the cells' heredity. Are they alive when inside the bacteria but dead when outside it?

Conversely, should we consider an ant colony, or the city of Calcutta, as living things? From a human-sized perspective, the individuals that make up those entities are more plausible units of life, and the larger systems are just loosely associated collections of these units. But this perspective may be misleading. An intelligent bacterium observing a human being, or any metazoan, might think of them as just loose associations of independent cells.

One common solution to this problem has been to use the

cell as the fundamental unit of life. For our purposes this solution is unsatisfactory because it might miss forms of extraterrestrial life that were not composed of discrete cells. The alternative is to associate life with the whole organism. This could cause difficulties if we encountered life forms whose division into organisms is less clear than that of Earthlife. Our potential problems are minimized when we identify life with the largest interdependent unit—the biosphere. We have seen how on Earth the various organisms in the biosphere are interlocked with one another in elaborate cycles involving matter and energy, just as our own cells are combined into a larger organism. We might even think of the whole Earthly biosphere as a living being, composed of innumerable separate organisms as well as some parts temporarily removed from the process of active life, such as the oxygen in the atmosphere. This suggestion has been made by the chemist James Lovelock and the biologist Lynn Margulis, who call the biophysical being Gaia after the Greek goddess of the Earth. For the purpose of our definition, it is not the notion of a biosphere as a living being that is essential, but the biosphere as a unit whose presence is a way to recognize life.

Some striking concepts emerge when we accept the idea that the biosphere is the fundamental unit of Earthlife. The history of Earthlife then becomes the tale of the continuous survival and evolution of the biosphere from its origin on the prebiotic Earth. Replication and subdivision into organisms and species have been strategies adopted by our own biosphere to ensure its own survival, but they need not be the methods used by an extraterrestrial biosphere. A biosphere that has not specialized into many quasi-independent living things is easy to imagine and presumably existed at an early stage in the evolution of Earth. On the other hand, it is hard to see how individual living things could evolve in the absence of a somewhat advanced biosphere. Such organisms might exist, however, either as the remnants of a larger biosphere that has been killed by a gross environmental change, or by having become separated from their original biosphere. We will see that the first situation might be the case on Mars at present, while the second would apply to some versions of panspermia, as well as to human expeditions to extraterrestrial environments.

One reason that the identification of the biosphere as the unit of life may seem strange to us is that the metabolism of the whole biosphere takes place on a time scale of thousands or millions of years, a very long time in human terms; whereas the living creatures familiar to us display characteristic behavior over a much shorter time scale. However, the same might not be true in other environments. For reasons we shall discuss later, the whole of an alternative biosphere might metabolize at a rate as fast or faster than that of familiar Earth organisms, while the metabolism of the individual components of that biosphere could be incomparably faster than anything we are familiar with.

The situation we might face in recognizing some extremely high rates of metabolism is analogous to the problem that an intelligent bacterium would have in recognizing a Sequoia tree as being alive. The tree changes so slowly on the time scale of the bacterium that it would appear to be a constant part of the environment. It would take a bacterial civilization that kept records over nine million generations (equal to perhaps fifty years) to notice some slow growth of the tree. There is some limit to how slow a living thing could metabolize and still have evolved and functioned in the actual Universe, (which as we know is only ten to twenty billion years old). But within this constraint there might be organisms whose metabolic times extend over thousands or millions of years, even if they are much smaller than Earth's biosphere.

We must now describe those features of the biosphere that identify it as alive and distinguish it from a non-biosphere. Consider, for example, all of the circulating water on Earth to be one entity—the hydrosphere. The hydrosphere, like the biosphere, is subdivided into a number of "species": rivers, ponds, icebergs, rain clouds, etc. These species interchange material in cycles; some of the steps require energy. Why do we consider the biosphere to be living and the hydrosphere nonliving (apart from the fact that the hydrosphere is mostly a part of the biosphere)?

We saw earlier that neither ordinary intuition nor standard biology furnished an answer that would be satisfactory for extrapolation to extraterrestrial life. We must therefore pursue the question a bit further, where we will still find that the two aspects of life, structural and functional, must both enter into a definition. Living things are made of the same type of matter—atoms and their combinations—as nonliving things, and the same general kinds of physical and chemical processes take place among the atoms in living things as in the nonliving. Therefore, a distinction and a definition cannot be found on the ultimate levels of structure and function either. The special features of life must be found somewhere in between, on the path between molecules and microbes; atoms and amoebas.

A plausible candidate for one of these features is the organization, or orderliness, of the atoms in living things compared with those of the nonliving. We believe that this is an important part of the definition of life. But the notion of orderliness is a subtle one, and we want to consider it separately before we discuss its role in life in particular.

ORDER

The word order is often used in everyday life in a way that does not apply in our discussion. Johnny's mother sees that his room is messy and tells him to put it in order. By that she means neat and tidy, everything in its place. We will apply the word ordered to situations which are unlikely to occur by random process. For example, suppose we made a row of a dozen symbols on a typewriter by pushing down at random any of the twenty-six lower-case letters or the spacer.

The arrangements "aaaaaaaaaaaa" and "bcbcbcbcbcbc" look very tidy, while "hppeanpsojoh" does not. In the way that we are using the terms, all three rows are equally orderly; each would occur about once in every 150,000,000,000,000,000 (1.5×10^{17}) attempts, if the typing were truly done at random. The appearance of any row that was specified in advance of its occurrence would be an indication that some nonrandom process was at work.

Yet, the first two rows immediately strike us as being special, while the last one does not. However, specialness is in the eye of the beholder; it is by our human choice that a particular row of letters appears that way. The string of twelve symbols that spells out "good morning" would also strike us as being special, but a Hungarian who knew no English might not agree. In fact, the third row of letters above (hppeanpsojoh) that looked so ordinary is actually "good morning" written in a simple code. Each letter was replaced by the next one in the alphabet, with "a" substituting for the space. We cannot know by inspection alone that any row of letters is not special in some way; we can in fact make it special by inventing our own code if we so choose. Furthermore, the rows of letters that appear special to us, such as "aaaa . . ." are often those that can most easily be produced by simple deviations from random processes, as for example if a child typed the same letter over and over. This kind of simplicity is not what we are after in defining life. For all these reasons, the criterion of appearing special is not a good one to use for orderliness.

Suppose, however, we find that ten rows of letters have been typed and that all are identical, whatever the message may be. This would indeed be an indication of some orderly, nonrandom process taking place, such as the control of the keyboard by a prearranged program. The order is connected with a whole collection of rows of letters, rather than the contents of any individual one, which may or may not appear special.

The amount of order present in the collection, in the

case of identical rows, can be measured by the number of rows multiplied by the number of characters in each row (ten times twelve). If the different rows of letters in the collection are not all the same, the amount of order is smaller, although the collection may still be more orderly than a random collection. There is a mathematical expression for determining the amount of order in any collection, whether of rows of letters or anything else. To apply this, we need to calculate the probability of obtaining the specified collection by random choice among all possible collections of that general type. The greater this probability, the smaller the amount of order in the collection.

The probability is influenced by the extent of similarity of the different rows in the collection and by the number of similar rows. In general, the amount of order will increase as the rows become more similar and as their number gets longer. The influence of the similarity of rows is a subtle one. If the number in the collection is kept fixed while the different rows become more similar, then the number of times each distinct row occurs must increase. It is this that decreases the probability of obtaining such a collection by random processes, and therefore increases the measure of order. The same conclusion is true for collections of objects other than rows of letters. For example, one reason that a book written in English is more orderly than a random collection of words is that many words in the book will be repeated several times, whereas the random collection is unlikely to contain any repetitions.

Returning to our typewriter example for a moment, we should point out that we set up the process so as to introduce some order in the collections regardless of the row of letters. We restricted ourselves to lower-case letters and the spacer. If we had permitted the use of the whole typewriter keyboard, including capitals, numbers, and punctuation marks, then it would have been very rare to find a collection of rows containing only lower-case letters and spaces. In other words, the rows described above all have a similarity that corresponds to some degree of order among them.

To estimate the order of a collection, we must know something about the probabilities of the individual members. For the rows of letters, we have taken all individual rows as equally probable. Suppose alternatively that the letters, instead of being chosen randomly, are chosen by a pattern that respects their frequency of occurrence in English. In this case some rows, containing more of the favored letters, would be more probable than others. Collections in which these rows appeared more often than others would not be more orderly than other collections. Order exists when a narrow selection has occurred among equally probable choices, not simply when more probable choices occur more often.

ORDER IN THE BIOSPHERE

Let us now extend our discussion of order to Earthlife. We can hypothetically take either the entire biosphere at a given instant of time or any portion of it that is readily accessible to us, such as a flask full of bacteria, and regard it as a collection of chemicals. (Scientists have not yet, in actual laboratory practice, been able to determine the total chemical content of any organism more complex than a small virus. We have learned enough, however, so that we can extrapolate to whole organisms and the biosphere.)

The whole biosphere, or even a small part of it, has several easily recognizable forms of order, all in a very high degree. There are many possible small or medium-sized organic molecules, stable under Earth conditions, containing up to a hundred atoms. If these molecules were being produced by random processes, we might expect all of the many types to be present among the 10^{43} molecules in the biosphere in roughly equal number. Instead the biosphere contains very large numbers of a few thousand different molecules and none (or very few) of millions of other molecules. Suppose we think of each molecule as a letter, chosen from several million possibilities. We sample the biosphere in many different places and times, in each instance making a list of the most common molecules present. This will produce in each case a sequence of letters. All these sequences will be very similar to one another, and the complete collection will therefore be extremely orderly. Because of the large number of molecules in the biosphere, its chemical content represents the greatest amount of order we know of.

We can recognize another type of order in the biosphere if we examine some of the large organic molecules present, such as nucleic acids and proteins. We find that these macromolecules are polymers of only a few types of the already restricted set of organic molecules present in the biosphere. The macromolecules are essentially natural letter sequences of the type we have discussed, but made up of a very few of the thousand or so letters that are in the biosphere pool. The proteins are all made of the twenty amino acids; the DNA molecules all contain the four bases A, G, C, and T; other natural macromolecules are similarly limited in their composition. The repetition of this pattern—as opposed to the occurrence in the biosphere of polymers made up of hundreds or thousands of different monomer units—again represents a high degree of order, although not as much as that present in the restriction of the biosphere to a few thousand types of molecules.

If our sample of the biosphere contained a multicelled organism or a colony of microorganisms of the same type, an additional dramatic type of order could be recognized. The longest molecules, DNA, that are present (in our analogy, the longest collection of letters) would be identical, or nearly so, both in length and in the entire sequence of bases, among the different cells in the sample. This is like getting many copies of the same long book. It is this form of order that is usually cited as characteristic of Earthlife, and it is indeed important. But it is numerically much smaller than the other types we have mentioned. The order represented by the exact sequence of bases in a bacterial strand of DNA is about the same as that in this book. It is certainly a lot of order but insignificant compared with that actually present in the complete DNA structure. This is true because the DNA contains only the bases A, C, G, and T rather than the many other small bases that it might contain, and also contains the same sugar unit at every location in its backbone.

To summarize, the most important type of order in the biosphere is that which results from the selective enhancement of a few types of organic molecules. Thus, it is distinguished from environments in which the synthesis of organic molecules has occurred (perhaps through Miller-Urey processes), but there has been no such selective enhancement of any type of molecule. The secondary order distinguishes the macromolecules in the biosphere from random polymers which might be produced as a mixture of the several thousand types of organic molecules found on Earth. Finally, the tertiary (but still great and important) order distinguishes the DNA of one living thing and its nonmutated descendants from that of any other living thing, or from the much greater class of DNA polymers which do not function as living things at all.

THE MAINTENANCE OF ORDER

We have examined the chemical composition of the biosphere at an arbitrary frozen instant in time. If we follow it over a period of time, we note that the order is maintained, or even increased, in opposition to natural forces which tend to dissipate it. A single bacterium, for example, can pass the order inherent in its strand of DNA to a descendant, or extend the order by giving nearly identical copies to many descendants. This survival of the nucleic acid through time is a remarkable phenomenon. For one thing, the individual bases continually tend to fall off the backbone of DNA. For the bacterium, *Escherichia coli*, that we met in our study of the architecture of life, this might occur to about one base every hour at our body temperature. The loss of a base is generally followed by a break in the DNA chain at that point. These processes, if unchecked, would gradually destroy the order of the DNA. During the life of the bacterium, these "errors" are repaired rapidly by specific enzymes. If the bacterium died, this order, and other types present, would tend to be lost, as the chemical contents of the bacterium started the trip back to equilibrium. Unless the dead bacterium were isolated from the biosphere by some means, it would not proceed very far on the trip. Its remains would be utilized by another organism, and the chemicals incorporated into some other ordered structure. Individuals perish, but the biosphere maintains itself.

For a time in the development of scientific thought, the occurrence and maintenance of this high degree of order in living things appeared to some as an indication that living things followed their own set of laws, unlike other objects in the Universe. We now understand that this is not so. Both living and nonliving matter obey the same set of physical rules.

THE RULES THAT GOVERN ORDER

The fate of order in the Universe is governed by the principle which we introduced in our chapter on energy, the second law of thermodynamics. In its simplest version, applicable only to isolated objects which exchange nothing with their environment, the principle states that order always decreases. When a part of the biosphere becomes isolated from its environment, as we mentioned in connection with our dead bacterium, its internal order does gradually decrease. But living things and the biosphere itself are not isolated from their environment because they exchange matter and energy with it. It is this exchange that allows living things to maintain and increase their order. The situation is like the poker game in Chapter 3, where a constant inflow of new money prevented it all from ending up in the kitty.

The rate at which order is lost is not mandated by the second law of thermodynamics. It depends on the detailed properties of the object, as well as on such factors as temperature. Bacterial spores can survive for a long time without external sources of matter or energy, as can tardigrade powder. Also, at the low temperature of liquid nitrogen ($-190\,°C$), all chemical reactions, including those that would disorder the nucleic acid chain, are slowed down tremendously, and cellular order can be maintained almost indefinitely.

The repair of damage to DNA in a living thing requires matter and energy from the environment. When an object does interact with its environment, its order can either increase or decrease. What happens is that the change in the order of the object has two sources: one, the processes occurring within the object itself, which decrease its order, and the other, a flow to or from the environment, which can increase or decrease the order. If the order of the object increases, then the order of the environment must decrease by an even greater amount. However, the environment is generally much larger than the object, so that a change in its order may be difficult to notice.

The simplest case beyond a truly isolated object is one that can exchange heat, but not any other form of energy, with an environment that is at the same temperature as itself. In this case what happens is that the object can undergo change only if it contains a supply of what physicists call free energy; that is, energy that could, under suitable conditions, do work on the environment. When change does occur, the order of the object may increase (although it need not) provided that heat flows out of it into the environment. The change can continue until all the free energy in the object is used up, at which point equilibrium is reached.

A simple example illustrates how an internal supply of

free energy can be used to increase the order in a biological system. If a bacterium is placed in a suitable nutrient solution containing organic chemicals and a few salts, it will grow and divide. The bacterium will convert a portion of the solution into new bacteria, and the remainder to waste products and to heat, which is absorbed by the environment. The chemical energy of the original nutrients has been used. A portion has been released as heat and as chemicals of lower energy, while another portion has been converted into the order present in additional bacteria. The system of bacteria plus nutrients has become more orderly, but has less free energy.

Another example of this process would be the development of a chick from a fertilized egg laid by a hen. The newly laid egg contains free energy and matter that is not very ordered, plus an ordered nucleic-acid strand. In the development of the chick, some of the free energy of the food in the egg becomes depleted and is used in the process by which the food gets converted into copies of the original nucleic-acid strand.

A simple nonliving analogue of this process is the growth of a salt crystal in a solution. During this process, the free energy that is present in the dissolved molecules decreases, while the number of molecules in the more orderly crystal-line state increases.

There is another way of looking at what happens during crystal growth or bacterial growth. We can think of the crystal, or the bacterium alone, as the system, and the medium as its environment. In this case the object is ingesting free energy from its environment, and uses it to increase its own internal order, eventually returning the energy as unavailable heat back to the environment. When the object uses up all of the free energy in itself and in its environment, the growth of order comes to an abrupt end. If the system-plus-environment is small, this usually happens quickly; and so for a crystal the growth of order is of short duration. For a large object, such as the whole Earth, the free-energy content may be great enough for order to increase for a relatively long time, but it still must inevitably end unless a new source of free energy is found. Something like this may have happened in the early stages of the biosphere to living things which lived on the large amounts of free energy that had been stored up in the prebiological stages of evolution. We have seen that eventually living organisms learned to use visible sunlight through photosynthesis as a source of energy, and so tapped a constant energy flow rather than exhaustible stored energy.

In a process like photosynthesis, what occurs is an increase in the order of the plant at the price of a decrease in the free energy of the environment. When this happens, it is not necessary that free energy be stored in the object over any length of time, although this sometimes does take place as a convenience. What is crucial is an inflow of energy in one form and its outflow in another form, in which it is less free, which allows order to be preserved and increased. The importance of a flow of free energy for the generation and maintenance of Earthlife has been emphasized by the biophysicist Harold Morowitz in his book *Energy Flow in Biology*.

A definition of life involving this behavior was suggested by the physicist Erwin Schrödinger in his influential book *What Is Life?* We think that this approach to the definition of life is an important part of the truth, but it must be extended if we are to distinguish life from other processes that increase order by ingesting free energy, such as crystal growth.

ORDER IN NONLIVING SYSTEMS

Living things are not the only objects in which interaction with the environment creates order in a system. We have already had the example of salt-crystal formation, which could occur naturally when the Sun shines on a pool of salt water. As time passes, the water evaporates, leaving salt crystals behind and fresh water vapor in the atmosphere. This situation is more ordered than the original one.

Order can also be reflected in many other ways. Many such types of order have been discussed by scientists studying nonliving systems. One type is based on orderly motion rather than orderly arrangements, and exists for example in the strangely behaving material known as superfluid helium. In this substance, which exists only at temperatures very near absolute zero (−273°C), the atoms move about freely as in a liquid. There is no order to their spatial locations, but there is a simple relation between the way that different atoms move. This relation is the same in any sample of superfluid helium, so that the motion is orderly in the sense that only a small fraction of all the possible types of motion are carried out by the atoms. This orderly motion is responsible for a strange kind of cooperative behavior by the whole liquid. For example, the helium can creep up the sides of an open container and spill over the top in a chilly parallel to the behavior of a swarm of ants in an open bottle. The orderly motion is all the more surprising in view of the fact that the atoms in the liquid are very close together and influence each other's motion. This motion is not like that of the planets about the Sun, where each planet is little affected by the presence of the others. Rather, in a superfluid it is the very presence of the many atoms that forces a common motion upon all of them.

Still another type of order involves directional properties of atoms, rather than location or motion. Many atoms act as tiny magnets with north and south poles. Ordinarily, in a piece of matter containing many atoms, these magnets point in random directions, and there is no overall magnetism. But in some substances like iron, it is possible to make most of the atomic magnets point in the same direction, and this results in the piece of iron as a whole acting as a magnet. Here the order consists in all the atomic magnets having a common direction, whatever their location in the iron. In fact, there are magnetic materials known as amorphous magnets, in which the magnetic atoms occupy random positions in the matter but the directions of the magnets are all the same. This situation is also maintained over time, even though the atoms are constantly affecting one another (Fig. 14).

Although various types of order exist on Earth, none except that of organic molecules have been known to de-

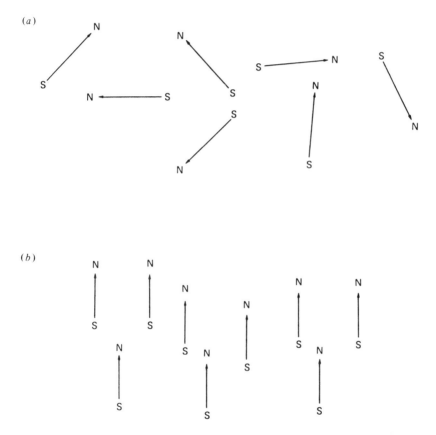

Figure 14. Disordered and ordered magnets. The directions of a disordered array of magnets point randomly, as shown in part (a) of the figure. When the magnets are aligned as by the action of a strong external magnetic force, these directions may all become the same, as in the orderly array shown in part (b).

velop sufficient complexity to display the richness of behavior we associate with living things. It is important in thinking about possible forms of extraterrestrial life to understand why this should be. One reason is because of the predominant types of energy flow that are available on Earth.

Not all energy flows through an object are capable of producing order. Sunlight that hits a charcoal block will be absorbed and heat the block slightly, but will not produce any order in it. Sunlight hitting a green plant produces order in the form of carbohydrate compounds, while sunlight hitting a human being produces at best a suntan.

An object exposed to an energy flow may react by increasing its own free energy, its temperature, or both. The actual result will depend on the nature of the energy flow and of the object. Only some systems respond by an increase in their order. On Earth the principal energy flow comes from both visible and ultraviolet light, and the most appropriate receptors are those in which chemical bonds can be made or broken. This is because the energy in one of these bonds is often about that of one photon of visible light. If the energy flow were mostly in the form of long-wavelength infrared or microwave radiation, the best receptors would be those in which the motion within a molecule was affected. This theme of coupling between energy and receptor will be important when we consider possible life forms in other parts of the Universe.

HOW ORDER BEGETS MORE ORDER

Our discussion of order would not be complete without some indication of an important aspect of the behavior of living things—their ability to create more life out of nonliving things. By ingesting such foods as glucose, which can be made by nonliving processes, a bacterium can make more bacteria. By eating a combination of foods, none of which need derive from a living thing, a human being can make more human cells. Since we identify life as a highly ordered arrangement of molecules, this amounts to saying that an ordered system can create further order by acting on a less ordered environment. Life is not alone in being able to do this. A small salt crystal placed in a saturated brine solution will cause some of the brine to leave the solution and enter the more ordered crystalline form. We have seen that such processes can occur only when there is a supply of free energy, either in the environment that is being transformed or flowing into it from outside. But now we are concerned with whether there are some general mechanisms, in life and otherwise, through which order is able to beget more order when a suitable environment is available.

In some cases the ordered system has a specific property as a result of its order, through which it is able to influence other, uncommitted units to partake of this order. For example, when many small magnets are pointed at random, the magnetic force they produce tends to be weak and hardly influences an extra magnet nearby. But if many magnets are in an orderly arrangement pointing the same way, they produce a strong magnetic force, which will act on a new magnet and influence it to point in the same direction as the rest, thus increasing the order.

A somewhat similar situation occurs in a laser. Here packets of light energy (photons) that are present in an enclosure stimulate the atoms also present to radiate still more light. If there are many different colors of light present, the atoms radiate randomly and produce a disorderly pattern of radiation. But if the photons originally present are predominantly of one color, the atoms radiate more of this color than of any other, further increasing the amount of order in the radiation pattern. We do not usually think of radiation as something that can be ordered, but according to the present scientific conception, it is as capable of order as is matter; and it is just when a radiation pattern is highly weighted in certain colors that it is orderly and will stimulate the emission of still more radiation of these colors (Fig. 15).

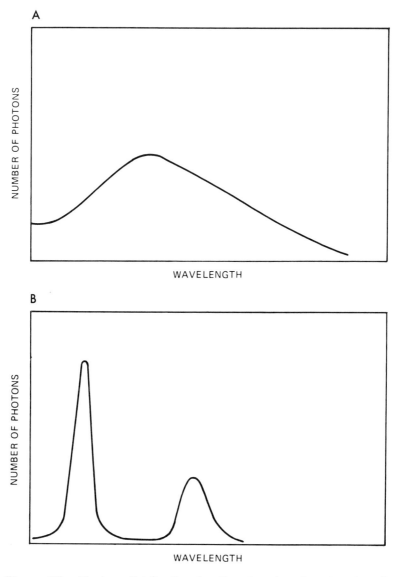

Figure 15. Photon distribution in disordered and ordered radiation. In disordered radiation such as that emitted by the Sun, there are many photons present for each of a wide range of wavelengths, as shown in part (*a*), with no great concentration at any wavelength. In ordered radiation, such as that emitted by a laser, the total number of photons may be the same in the first case, but these photons are concentrated at only a few wavelengths as shown in part (*b*).

Both of these mechanisms can be considered analogues of what the Earth's biosphere does when it increases its chemical order by preferentially synthesizing more of the biochemicals that compose it. The biosphere has a number of different ways of doing this, several of which we have described in Chapter 2. Many of these mechanisms do not directly involve the most celebrated of the ways that Earthlife uses to increase order, the replication of the nucleic acid strand. Instead they involve the selective production of a substance by the catalytic action of other substances. A strong case can be made for the view that it is this type of catalysis that has been the main thread in the long evolution of matter on Earth, from its prebiological beginnings through the origin of the first individual living things up to the present. According to this point of view, nucleic acid is a blueprint used by protein enzymes to make more of themselves rather than the other way around, and replication of nucleic acid is a particular step in the process of making more protein rather than the driving force of life.

Whatever the merit of this view, replication is an important process for increasing order. In Earthlife there is a fairly unique method for achieving replication which is based on specific geometrical arrangements of molecules that are favored by the electrical forces between various parts of the molecule and between one molecule and another. Because of these favored arrangements, certain enzyme catalysts, together with a single DNA strand, in the presence of unattached nucleic acid components build up a second DNA strand that is complementary to the first and attached to it. This second strand can then be used to repeat the process, forming yet another strand identical in its base sequence to the original strand, and therefore achieving replication of this strand.

Arguments have been given about why this rather cumbersome method of replication should have evolved. These arguments rest mainly on the need for faithfulness of the replication. It is certainly not the only method that can be imagined, nor is it clear that it would be an optimal method under different environmental conditions. A method of replication that involves only a single step, rather than two steps, is imaginable, even within the confines of chemical arrangements, providing that the object to be replicated is built of components that selectively attract identical components instead of complementary ones, as happens in DNA replication. We can think of DNA replication as a sort of mold and clay arrangement, with one set of bases as the mold and the complementary set as the clay that fits the mold (Fig. 16). Once the clay has set, it can be used to create yet another mold by pouring soft rubber around it.

We can also imagine pouring the soft rubber directly onto the opposite side of the original mold, thereby creating a replica of it. This would be an example of one-step replication. More realistic examples may occur in living things in other environments. Many atoms act as small magnets with distinct north and south poles. At Earth temperatures the magnets are usually not lined up in any special way because the collision with other atoms would rapidly disarrange them. At very low temperatures, this may not be the case, and an arrangement of atomic magnets may persist indefinitely. Suppose some arrangement exists such as the chain pictured in Fig. 17. When a new atomic magnet approaches the arranged group of atoms, it is more likely to align itself in the same direction as the nearest magnet it approaches than in the opposite direction. A series of such steps would lead to a second chain, whose arrangement of magnets would be the same as the original set. Therefore, the original set would have replicated itself, in the sense that an identical arrangement of magnets has been formed.

In this example the order that is being replicated is the arrangement of directions of the magnet along the chain. The chemical nature of the atoms that comprise the magnets is less essential, although it might play some role in determining the rate of replication. This suggests an important generalization about the role of replication in life. The only

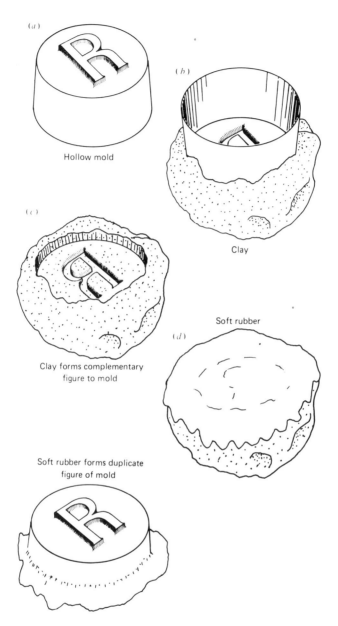

(a)

Hollow mold

(b)

Clay

(c)

Clay forms complementary
figure to mold

Soft rubber

(d)

Soft rubber forms duplicate
figure of mold

Figure 16. A mold, clay, and rubber analogue for DNA replication. The hollow mold in (a) represents an original DNA strand.
The clay in (c) that has had the pattern impressed into it represents the complementary strand. The clay then acts as a new
mold for the soft rubber, producing in it a pattern duplicating the
original mold.

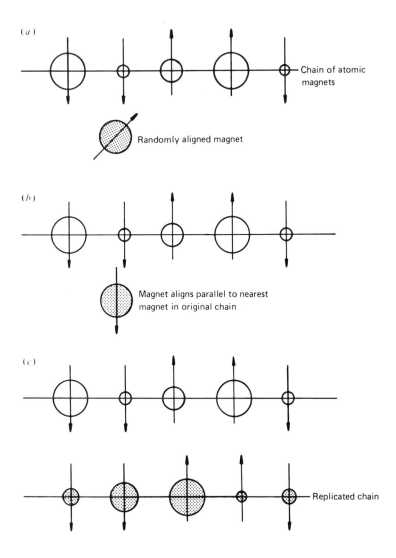

Figure 17. The hypothetical replication of an ordered chain of magnets. A randomly directed magnet approaching the chain will line up with its direction parallel to that of the nearest member of the chain. When this occurs for many magnets, a new chain is eventually formed that duplicates the original chain in the arrangement of directions of its magnets. This method of replication might become realistic if some mechanism existed for keeping the magnets in place along the chain, as the backbone does for the bases in nucleic acid.

aspects of an object that need be replicated accurately are those that are involved in the process itself. Other aspects may be incidental to the replication, and these may not be replicated accurately. As an example, it is known that many atoms occur in several different forms, known as isotopes, in which the atomic nuclei have slightly different weight. However, the isotopic form of the atoms in a nucleic acid molecule plays little or no role in the replication process. We would therefore not expect that a replicated strand would be isotopically identical to the original strand. The features that are essential for nucleic acid replication are the chemical properties of the bases and they are accurately replicated. But in another situation, the isotopic composition might be the essential feature that allows for replication, and the chemical properties of the atoms unimportant. In the latter situation, we would expect accurate replication of the isotopic content, but little regard to the chemical structure.

Factors such as this may influence how complex a system must be before it can begin to replicate itself. We have seen that the nucleic acids that exist are much too complex to have originated as random fluctuations, and that if they were the first replicating systems, an immense amount of prebiological evolution had to occur in order to produce them. In other circumstances, there may be replicating systems that are much less orderly than the nucleic acids. If so, these systems might arise as random fluctuations in an environment and then, through their replication, lead to a great increase in the overall order of the environment. In other words, while it appears likely that on Earth the order of chemical specificity preceded the spatial order of nucleic acid chains, in other environments an analogue to the spatial order may have occurred first and, through its replication, have led to the order of specific chemical or physical combinations.

Another process by which order can come from disorder is known as nucleation. Even when the formation of an ordered situation is energetically favored, such as the freezing of a liquid cooled below its freezing point, a series of steps is necessary by which the atoms of the liquid, which are relatively disordered, go into the orderly spatial arrangement that they have in a solid. The method by which this occurs, which is also relevant to some other cases, involves the production, by random fluctuations, of small amounts of the ordered situation. These centers of order act as nuclei from which more order can grow. These nuclei are unrelated to cellular or atomic nuclei. The word just refers to the place where a process starts. The nuclei can promote order because if one is big enough, an individual atom nearby will find it energetically favorable to line up with the other ordered atoms rather than maintaining its free unordered state. The situation is similar to a small country choosing to unite with a much larger one in order to benefit by increased trade, more security, etc., rather than taking its chances by remaining independent. The nuclei must be reasonably large for this to work, otherwise there is too great a probability that the nucleus will dissipate before growing to a self-sustaining size. Often there will be many nuclei competing simultaneously for the disordered atoms, and eventually these extended nuclei must themselves unite to make the final ordered situation.

In the context of life, one could regard a living thing, such as a bacterium in a nutrient solution, as a nucleus that triggers the change of the material in the solution into an ordered state. This process is aided by the fact that as the bacterium divides, it produces yet more nuclei which can, by moving away, stimulate further changes in regions that are far from the original bacterium. So, nucleation is an important part of the life processes of Earth's biosphere, although it does not play any obvious role in those of individual living things.

We have mentioned a few of the methods by which order can emerge from chaos if the overall conditions are ripe. The ones we have cited do not exhaust all that are known. Furthermore, there are processes, such as the ordering of motion of helium atoms that takes place at low temperatures, for which the mechanism of ordering is not well understood. The variety of mechanisms available suggests that self-replication, while it is a ubiquitous ordering process in Earthlife, is by no means a universal necessity for producing order. We must allow for the possibility that in other situations, life may find it more convenient to base its generation of order on other methods, just as presumably happened on Earth during the period of chemical evolution that preceded the first individual living things.

THE DIFFERENCE BETWEEN LIFE AND NONLIFE

We have now seen some of the ways order can be induced in a number of systems when free energy is available. What then is the fundamental difference between what happens in the biosphere and what happens in other, nonliving systems?

We believe that it is through quantitative differences, rather than through some special qualities, that a distinction between living and nonliving matter should be made. Living things do not differ in kind from some of their nonliving analogues that we have cited, but rather in the extent to which they carry out the activity they both share—the creation and preservation of order. Among the organisms now on Earth, this quantitative difference is so marked as to create a gross dissimilarity in behavior between living and nonliving, which has led to the distinction in the way we talk about them. In earlier stages of Earth's history, before Darwinian evolution began its long sharpening of differences in behavior, there was probably much more continuity between living and nonliving things. It is also plausible that in other environments, objects exist whose capability for the creation and preservation of order is intermediate between those things we call living and those we term nonliving.

Furthermore, there is little reason to think that Earthlife represents the ultimate possibility for how matter can become ordered through energy flow. We can imagine other environments, or perhaps even the future environment of Earth, in which the distinction between living and nonliving is carried still further than now. The living might share fewer of the characteristics of the nonliving than is now the case on Earth, even to being composed of different types of matter or based on aspects of physical law that are unessential to the nonliving parts of the Universe. It is not only through the discovery of creatures of superior intelligence that extraterrestrial life may humble us. We may also find that Earthlife has far to go in its expression of the possibilities inherent in the universal properties of matter and energy. Our extraterrestrial cousins may be much farther down that road, toward whatever maximum order matter and energy can produce.

CYCLES AND CATALYSTS

Why do living things have a greater capacity for producing order than nonliving ones? The reason seems to be that living creatures use a complicated cycle of processes to convert the matter and energy they ingest from the environment into a form which adds order to them. We have seen that in all living cells, thousands of chemical and physical processes are going on in a carefully arranged sequence of steps (Fig. 18). Such biochemical processes often involve cycles, in which a substance produced in one step combines with another

Ornithine + Carbamoyl Phosphate $\longrightarrow$ Citrulline + Phosphate

Citrulline + Ammonia $\xrightarrow{\text{Several Steps}}$ Arginine

Arginine $\longrightarrow$ Ornithine + Urea

Figure 18. A typical biochemical cycle. Its overall effect is to convert two substances, carbamoyl phosphate and ammonia, into two others, urea and phosphate. The various steps are catalyzed by enzymes. The cycle has been greatly simplified for clarity.

172

produced in a second step to produce a new one which will be used elsewhere in a later step. It is the existence of this "dance of life" that makes some biologists prone to teleology, the view that living things have some built-in purpose, or to a belief in the special creation of living things. But other scientists have recognized that biochemical cycles (although much more complex) are similar in kind to processes occurring in systems which we do not consider alive. For example, we might compare a simple case of a cellular cycle with a cycle which occurs in the upper atmosphere to produce and destroy ozone, a form of oxygen that strongly absorbs some wavelengths of ultraviolet light (Fig. 19).

$$O_2 + \text{Ultraviolet Photon} \longrightarrow O + O$$

$$O + O_2 + M(\text{Catalyst Molecule}) \longrightarrow O_3(\text{Ozone}) + M$$

$$O_3 + \text{Ultraviolet Photon} \longrightarrow O_2 + O$$

$$O + O_3 \longrightarrow 2O_2$$

Figure 19. A cycle of reactions by which ozone is produced and destroyed in Earth's atmosphere. The amount of ozone present in the atmosphere depends on the intensity of ultraviolet light and on the rate of the various reactions.

It is the biochemical cycles in living things that make them as adaptable as they are and able to use a variety of forms of matter and energy which may become available to them. A crystal of sodium chloride will only continue to grow as long as sodium and chlorine ions (or ions of a very similar size and charge) are available to it, even if it is immersed in a concentrated solution of other atoms and ions. On the other hand, a bacterium can process a large number of organic chemical compounds into the same end product— more bacteria. The reason is that among the many cycles that can take place in a bacterium, one or more can be found that use a given organic compound as one of their working substances, and when this compound is found in the bacterium's environment, this is the cycle that operates. Of course, even bacterial cycles are not omnivorous, as can be seen from the piles of discarded plastic objects that fill our garbage heaps.

The existence of effective linked cycles in living things must be the result of the history and evolution of life. We have seen that a great deal of prebiological evolution was necessary to account for the large amount of order inherent in the high concentrations of some organic molecules already present on Earth before any living things had developed. This evolution may have originally started with relatively simple cycles of the type found outside living things. These cycles can lead to the preferential production of some simple compounds, those which are part of a cycle, over other substances not involved in a cycle. Such cycles are a generalization of the idea of a catalyst, a substance that promotes the formation or destruction of other substances without itself being consumed in the process. Individual catalysts, the enzymes, play an essential role in many biochemical processes, including the formation of nucleic acid strands from monomers. What we are now describing is a situation in which not an individual substance, but a whole cycle of processes involving many substances, acts as a catalyst for the production of some materials.

The chemist Manfred Eigen has described the possibility that a number of such catalytic cycles might become linked up into what he calls a "hypercycle." The product that is catalyzed by one cycle takes part in another cycle, leading to still greater concentrations of the substances involved in all the cycles. Eigen has argued that such linked cycles led to the high degree of chemical order that now characterizes the biosphere.

Just as some of the physiological functions of higher species are hardly different from those of the primitive organisms from which they evolved, so it is plausible that many

174

of the same cycles that eventually led to the first living things should still play a role in the way they function today. The information needed to ensure that the cycles in living things on Earth work properly is stored in the sequence of bases in their nucleic acids. Thus, there is a close connection between the way in which order is stored and the way in which it is maintained and increased. This feature is much more pronounced in living organisms than in some nonliving ordered systems such as crystals, where the vehicle for storing order plays only a minor role in the generation of more order. Through catalytic cycles it is possible for a relatively small amount of order contained in the nucleic acids to produce a significantly greater amount of order, as for example when a bacterium grows and divides in a nutrient medium.

Cycles play other roles in the metabolism of the biosphere. We have mentioned some of the grander cycles in which the individual elements are not molecules but groups of living things, such as plants or animals, which serve to maintain the concentration of some substances important in the biosphere, such as free oxygen. According to one point of view, nucleic acids and the organisms containing them are a strategy through which the biosphere maintains and increases its chemical order. Even though some of the organisms have developed their own views of their purpose, they still serve this primordial function of helping to slowly convert chaos into chemical order.

ALTERNATIVE BIOSPHERES

We can now return to our general definition: Life is fundamentally the activity of a biosphere. A biosphere is a highly ordered system of matter and energy characterized by complex cycles that maintain or gradually increase the order of the system through exchange of energy with the environment. Earthlife is one specific expression of this theme. What variations on it may exist elsewhere?

On Earth a prominent feature of the biosphere is the

presence of individual living things. Each undergoes the process of active life, which also involves the exchange of free energy with the environment to maintain the order of the individual. The genetic order contained in these organisms is a kind of blueprint for the cycles by which the biosphere gradually evolved. It is the close relation between order and its expression in cycles that distinguishes living creatures from other ordered objects. In other places in the Universe, biospheres may exist which have as high an order as Earth's biosphere but do not contain individual units of high order.

Evolution might have taken an alternative path in which individual living things did not develop. It is also possible that some biospheres which have once contained individuals have evolved to the point where the individuals have coalesced and are not really independent. In a future biological state, individuals may be as closely knit in their functions as our body cells are today.

Replication is another feature of Earthlife that might not be present in alternative biospheres. Even nonfaithful reproduction might not exist. We have seen that other mechanisms for creating order do exist. Replication is a strategy for increasing order, possessing the important advantage that it avoids the problem of maintaining order within a large volume with a small surface area. Perhaps this is one reason why replication arose in the first place. If a living thing becomes too big, it may not be able to ingest free energy and excrete heat quickly enough to compensate for the disorder produced by its activity. This is another example of the square-cube law.

The activity of a living thing occurs over its whole body, while ingestion takes place on the surface. As the size increases, the ratio of surface to volume decreases. This problem can be solved in several ways. One is for the organism to develop many zigzags on its surface and to ingest or excrete through all of them. Another is for it to divide into two smaller versions, each of which will have a surface that is relatively larger compared to its volume than the original. The two smaller living things can then move apart, if only by random environmental disturbances, and for a while each will avoid the problem of the small surface-to-volume ratio that bothered their parent. Thus, replication may have arisen as a way of avoiding an internal pollution crisis in very early living things. Note that the entire biosphere has not needed to replicate itself thus far. One reason is that while the biosphere has grown over time, it has remained two-dimensional, essentially confined to the Earth's surface. Because of this, its volume and surface have remained approximately proportional.

Replication also has its disadvantages. The replicated form must compete with its copy for the energy and materials that each needs to maintain order. This is not a serious problem when life is sparse and energy and materials abundant, but it becomes serious as life flourishes. Earthlife has dealt with this problem by evolving into newer forms which could use new sources of energy and matter. Viewed abstractly, this whole procedure appears a bit like burning your house down in order to convince yourself to move elsewhere. It is what we have, but we shouldn't jump to the conclusion that the same procedure will happen elsewhere.

In an alternative biosphere, the possible range of variation in living things might be much smaller because the physical laws by which they operate constrain their form much more than in our biosphere. For example, the equivalent of the genetic code might allow for the synthesis of only a small number of different catalysts. In such a situation, living things would automatically be very similar to each other, not because they are all replicas of a common ancestor but because only a few types could arise at all. This kind of life would perhaps be less interesting to a human observer than the rich variety of life forms on Earth, but it need not be less efficient at creating order, especially in an almost constant environment.

One alternative to replication might become the rule in a

mature biosphere when there were no new "ecological niches" to be filled. This alternative involves a close symbiosis among a very large number of distinct individuals, each of which performs a different function in the overall operation of the biosphere. Replication of any of the individuals, except for replacements, would not improve the situation, but rather would result in competition for the available nutrients. In this particular case, there would be selective pressure against replication. In the parts of the biosphere where such replication occurred, there would be a higher probability of disrupting the complex cycles which maintain order in the region, and so a higher probability that all the living things in the region would die. A situation like this seems to occur in adult multicellular organisms, where uncontrolled growth of some cells at the expense of others must be avoided.

The definition of life that we propose has several consequences for the detection of extraterrestrial forms. The most obvious is that we must consider possible biospheres in other environments, instead of just other individual creatures that may be found there. In some sense this makes the search for extraterrestrial life easier, because we can imagine detecting some aspects of a biosphere at a distance more easily then we could detect individual living things. James Lovelock has pointed out that the existence of certain combinations of gases in the Earth's atmosphere (such as oxygen and methane) which cannot exist together in equilibrium might be taken as a sign that our atmosphere is a part of complex biospherical cycles that continuously produce and destroy these gases. A problem with using such tests for extraterrestrial life is that some similar departures from equilibrium can occur as a result of nonbiological cycles generated by energy flows of a type that are known to occur in Earth's atmosphere.

In the following chapters, we will apply our definition of life to the questions of how it may have evolved in other environments in the Universe and in what forms it may exist there. A crucial question in this regard is what constraints exist on life in various environments because of the general laws of physics and chemistry. We will turn to a discussion of these constraints in the next chapter.

Chapter 7

Conditions for Life

On the basis of our definition of life, we can identify several crucial requirements that must be met for any life to exist. There must be a flow of free energy, a system of matter capable of interacting with the energy and using it to become ordered, and enough time to build up the complexity of order that we associate with life. When these necessities are supplied at a particular place in the Universe, however exotic and unearthly that location may seem, it is where life may be found.

Many writers have taken a much more restricted view of the conditions for life. They have generally assumed that it will only flourish under conditions suitable for Earthlife. Proponents of this view have included William Whewell more than a century ago, Lawrence J. Henderson early in this century, and George Wald more recently. They have held that the presence of liquid water, a certain temperature range, and other specific environmental factors were indispensable for life. We think that this narrow Earth-centered viewpoint misses many other possibilities. In this chapter we will examine the basic constraints on life from the viewpoint of the general laws of science, rather than Earth-based prejudices. Later we will explore the various environments that meet the necessary conditions and speculate about what creatures may live there.

FREE-ENERGY SOURCES FOR LIFE

Two principal requirements for life are a source of free energy and a "sink," or place to which the energy can be transferred in a degraded, less-available form such as heat. A sink for the degraded energy is necessary because the maintenance of order requires an outflow of heat. If an ordered living thing and the heat it produces are not soon separated, the order will eventually be destroyed. Often the sink is a place which is at a lower temperature than the object, so that heat will flow to it. For example, Earth's biosphere, which is at a temperature of about 20°C, emits its degraded energy to outer space at a temperature of −270°C.

A variety of energy sources exist in the Universe and can be used as fuels for life. Energy flows such as sunlight can be tapped directly, or energy stored from past flows, such as the chemical energy in an ear of corn, can be eaten. We have mentioned the various forms of electromagnetic radiation, such as radio waves; microwaves; infrared, visible, and ultraviolet light; and X-rays. Each form represents an important energy flow in some environment. Another important type of energy flow is streams of charged particles. Stored energies also occur in several forms. Energy is stored in various environments as a difference in temperature between places, as the chemical energy of tightly bound atoms, or as the nuclear energy of certain elements.

If a type of free energy is to be useful to a specific form of life, the energy must interact with the matter composing the life. For the living things of Earth, light energy and chemical energy are the useful types, as they can most easily make specific changes in the molecular combinations used to store order. These types of energy are also the most plentiful on Earth. Other types of free energy, such as X-rays, do not couple easily to the formation of molecules and so would at best be a nuisance for Earth-type chemical life, even if they were more prevalent here. But a life form that stored order through the capture and loss of tightly bound electrons in heavy atoms might find X-rays very useful as a source of free energy, since X-rays are effective at detaching tightly bound electrons from atoms.

It is possible to make use of types of free energy that are not directly coupled to the mechanism for storing order, but to do so, a series of steps would be necessary. For example, if the packages of energy are larger than what are needed to affect the order-storing mechanism, they might be broken down into smaller packages. In the opposite case, several small packages would have to be combined into a single one. While both of these processes are possible and occur in some Earth organisms, this is a more sophisticated use of energy than the direct use of the primary energy, and therefore can be expected to occur later in evolution. Nevertheless, this possibility is important because we cannot expect that there will be a perfect match in each environment between the forms of matter suitable for organization and the types of free-energy flow available. The extent to which the matter and energy in some environments are "well coupled" will be an important factor in fixing the third general requirement for life—sufficient time for it to develop from nonliving things.

Useful flows of energy are most common at the interface between two very dissimilar environments. For example, much more energy flows inward toward the surface of the Earth than flows in either direction one hundred meters below the surface. One reason is that many forms of energy are quickly absorbed when they impinge upon a new medium. The starlight that travels across the Universe to us from a distant galaxy is absorbed by a millimeter of opaque paper. Only neutrinos, among all the forms of energy flow that we know, are almost unaffected by any matter in their path, and this very disdain for matter disqualifies them as an energy source. Because energy flows are often greatest at interfaces, the life that uses such flows will be a surface phenomenon, as is Earthlife. When matter is so rarefied that it does not impede energy flow very much life can feed off the energy flowing through a whole volume of space, rather than only that on the surface. This condition is not required for life that feeds off stored energy which can be distributed over a large volume, as we will soon see.

Interfaces are favorable locations for life for yet another reason. A flow of energy is useful only if it is not in equilibrium with the local material. For example, most of the radiant energy in the interior of a star is not useful as a source of free energy within the star because the matter contained there is at the same temperature as the radiation and so cannot gain or lose free energy from it. This is not true at the surface of the star, where the matter has access in one direction to empty space, which is at a much lower temperature. Even within a fairly homogeneous environment, it is possible for places far apart to be out of equilibrium with each other. An example is the increase of temperature as one travels into the Earth. Such variations can also be a source of free energy.

HOW MUCH ENERGY IS NEEDED?

In order to evaluate the amounts of free energy available in various environments, we can compare it to the rate at which life uses free energy. Of course, we would like to be able to know this for the actual form of life in each environment, but we don't have the information yet. The rate at which free energy must be supplied will always depend on how much order is involved and on how quickly the order would decay in the absence of an energy supply. We can get some idea of the numbers involved from a study of energy use by Earthlife. A large animal, such as a human being, must ingest energy at the rate of about one-half calorie per second per kilogram of its weight in order to function. (The calorie we are using as a unit here is the physicists' calorie, the amount of energy needed to raise the temperature of one gram of water by one degree centigrade. The Calorie used in diet books is one thousand times greater; physicists call it the kilocalorie.) Most of the energy ingested by a human being is eventually released as heat. The total emission rate corresponds to about the same output as a typical light bulb, or a hundred watts. The waste energy and other wastes we emit usually contain much less free energy than the food we eat, so the value of one-half calorie per second per kilogram can be used as an estimate of the free energy required for human metabolism. Other mammals have similar free-energy requirements. About the same value represents the rate at which the leafy part of a plant uses sunlight during photosynthesis.

On the other hand, much higher values apply for small living things like bacteria. The heat produced by a bacterium during growth, which must be similar to its ingestion of free energy, is at a rate of a thousand calories per second per kilogram, or a thousand times greater than for a large animal. An important reason for this large disparity is that the bacterium has a much greater surface in comparison to its mass than does the large animal. As a result, the energy flow through the bacterium can be much greater because energy must enter and leave through the surface. This is another example of the square-cube law.

The rate of free-energy input into a bacterium is about enough to produce a chemical reaction in each molecule of the bacterium once in a few hundred seconds, approximately the time it takes the bacterium to divide. The corresponding rates for human beings are a thousand times smaller, in agreement with the much longer reproduction times we enjoy. It is probably not an accident that the reproduction rate adjusts itself to the rate of energy intake, as the latter is much less under the control of evolution than the former.

This analysis suggests that a flow of free energy can be usefully described by how much order it can produce in a given time period for each ordered unit of the system absorbing it. In general, this quantity will depend on the details of the particular reactions involved in extending the order of the system. The less the energy needed to drive a reaction, the less the need for external free energy. If the free energy comes in separate packets, the way radiation does, and if one or a small number of packets are used to change a single unit of order, then the relevant measure will be the number of energy packets absorbed per second. Two situations in which this number is the same will be equally effective in producing order, whatever the size of the individual packets.

Another quantity of some interest is the total rate at which energy flows through Earth's biosphere. This has been estimated by Harold Morowitz to be about 10^{13} calories per second. This is about one-tenth of a percent (10^{-3}) of the total solar energy reaching Earth's surface, so that Earthlife is using only a small part of the available energy flow. If we divide this rate of energy use by the total mass of living matter in the biosphere (Table 3-1), it corresponds to a rate of about .01 calorie per second per kilogram. This is much less than the rate found for individual living things, such as large plants or animals, and thousands of times less than that for bacteria.

There are several interesting conclusions that we can draw from this difference. One is that the mass of bacteria growing or dividing at any time is very small compared with the total mass of living things. Furthermore, much of the total mass in living things is composed of supporting material, such as the trunks of trees, which does not metabolize at all. The typical atom in a living thing must wait for years before it gets enough energy to make or break a strong chemical bond. Yet the supply of energy is enough to fuel all the familiar activities of our biosphere.

With this guidance from Earthlife about what constitutes an interesting amount of energy flow, let us now go on a survey tour of the types of energy flow that are available in the Universe.

SOME NOTEWORTHY ENERGY FLOWS

The most widely distributed form of energy in the Universe is one that is not useful for the purposes of life. The whole Universe is filled with microwave radiation. This radiation was discovered in the 1960's by radioastronomers who detected the faint whispers of it that reach Earth. Most scientists believe that this energy is left over from the intense radiation which filled the Universe right after its origin some ten to twenty billion years ago. The energy flow from this microwave background is about 10^{-6} calories per second on a surface of one square meter, less than one hundred-millionth of the intensity of summer sunlight on Earth. But since this radiation is everywhere, the total amount is huge, equivalent to the light radiated by all the stars during the history of the Universe. Equal amounts of radiation flow in any direction at any time, unlike sunlight on Earth which only comes from a single direction at any given time.

One effect of the microwave radiation is to keep the Universe at a temperature of three degrees above the lowest temperature possible (absolute zero). The Kelvin temperature scale uses absolute zero ($-273\,°C$) as its zero point, with the size of the degree the same as in the centigrade scale. Zero degrees centigrade is thus $273\,°K$. For very low temperatures, we will cite both scales. An object in space, therefore, cannot have a temperature below $3\,°K$ ($-270\,°C$) for any length of time. If it is artificially cooled to a lower temperature, it will heat up to $3\,°K$ by absorbing the microwave radiation. This temperature of $3\,°K$ sets a condition under which a body can emit energy to the environment. If the body is at a higher temperature, it can do so. Most of the environments that we will consider as possible abodes for life are of significantly higher temperatures than $3\,°K$, so that they are able to emit energy to space. It is space at $3\,°K$ that acts as the ultimate sink for the degraded energy of all objects in the Universe.

The universal microwave radiation cannot be used as a

source of free energy for life any more than the heat energy in a large ocean at uniform temperature could be used by a fish swimming in it. There is other radiation in space, which has been produced by the stars during their long shining career. The ultimate source of this radiation is nuclear fusion deep within the stars. This starlight, unlike 3°K microwave radiation, varies greatly in amount and composition from place to place, and in some of these places it could be used as an energy source for life. This can happen wherever the radiation present in some region differs from the microwave background in any respect, such as direction, distribution among wavelengths, or intensity. When such a difference exists, the radiation can be absorbed in one form and emitted in another form containing less free energy. The difference in free energy is then available to create order. Over the whole Universe there is less stellar radiation energy than there is microwave energy. However, within the great collections of stars in galaxies, the stellar radiation is comparable in amount to the microwave background, and in the center of a galaxy or in the vicinity of a star, such radiation is much more intense.

The radiation originating in stars is mostly visible light or the similar ultraviolet and infrared light. Only a small amount of it is in the form of X-rays, which come in much larger packets. The ultraviolet and X-radiations are more important in some respects than their intensities would suggest because they are able to remove electrons from atoms, leaving behind positive ions, and to break up molecules into their component atoms, whereas the packets of energy in visible light, infrared light, and radio waves cannot do this. In the region of space near hot stars, the ultraviolet radiation is intense, and any atoms present tend to have at least some of their electrons detached. Also, molecules are not usually present because they are broken up into their component atoms. Sometimes this can be avoided if the density of matter present is high enough, because then the atoms first exposed to the ultraviolet radiation absorb it, shading the more distant atoms from its effects. This is similar to what happens on Earth, where ozone molecules in the upper atmosphere absorb ultraviolet radiation from the Sun, shielding living things on the surface. This shielding does not happen in parts of space where the ultraviolet light is most intense, but it does occur in other regions where there are dense clouds of atoms and dust. In these regions, which contain much of the matter in the Universe, the intensity of ultraviolet radiation drops off considerably, and neutral atoms and molecules can exist. These clouds will play a prominent role in one of our later scenarios for extraterrestrial life.

Another energy flow that may be used by life in some environments is the streams of charged particles called cosmic rays. Like X-rays, cosmic rays have the ability to displace electrons from atoms, producing ions. This is possible because cosmic rays, like X-rays, have sufficient energy in each individual packet to do this. While the energy of the individual particles in cosmic rays is greater in magnitude than that of visible starlight, the total energy flux from both sources is about the same, because the number of individual cosmic-ray particles is much less than the number of photons of starlight.

A cosmic-ray particle can ionize a number of atoms without losing much of its own energy. Ions are extremely reactive species, so the use of cosmic rays as an energy source for life would seem promising. However, in most places the number of cosmic-ray particles is too small to be of use as the primary source of free energy for a life process on a time scale of interest to us. The average atom in space will be ionized by a cosmic-ray particle only about once in three billion years, or just a few times during the life of the Universe.

In some specialized environments, however, cosmic rays may play a role in life processes. They penetrate the thickest galactic dust clouds and produce ions there. The molecules that are formed as a result of this process may play a role in another life form that we will discuss in Chapter 12. Cosmic rays can also become trapped by a combination of gravity and magnetic forces and concentrate in certain localities. Something like this has happened in the Van Allen radiation belts which surround the Earth at a height of several thousand kilometers.

In addition to the flows of radiation and charged particles discussed so far, there are other energy flows in various parts of the Universe. Some stars emit large amounts of radio waves, similar to those broadcast by radio and television stations on Earth. In most of the Universe, the total energy flux in these radio waves is small compared with that in other kinds of radiation. But because the energy packets of radio waves are much smaller than those, say, of visible light, the number of energy packets in the flux of radio waves is comparable to that in the flux of visible light. This means that if a suitable vehicle for storing order exists which couples well to individual energy packets of radio waves, there can be sufficient amounts of radio waves to serve as a free-energy source for this vehicle. Actually, radio waves of the right wavelength can be much more effectively absorbed in certain molecules than visible light or ultraviolet light, and a very thin layer of molecules may be enough to utilize all the energy in a stream of radio waves. There are also some stars and collections of stars that are especially strong emitters of radio waves. In the vicinity of these stars, the flux of radio waves can be more intense, in numbers of radio photons, than that of sunlight on the Earth.

A general conclusion to be drawn from this data is that in different environments there are a variety of energy flows which could act as free-energy sources for living processes. Some of these energy flows involve a rather small amount of total energy, but in terms of the number of packets that flow through any area per second, they compare favorably with the amount of sunlight reaching the Earth. It appears that most environments we know cannot be ruled out as possible homes for life on the basis of a lack of available free-energy

Table 7-1

ENERGY FLOWS IN THE UNIVERSE

Environment	Type of Energy Flow	Total Amt. of Energy Flow (calories per sec per cm²)	No. of Energy Packets in Flow (packets per sec per cm²)
Interstellar space (typical place in galaxy)	Visible light	10^{-9}	10^{10}
	Cosmic radiation	10^{-9}	10
	Ultraviolet light	3×10^{-13}	10^6
Interstellar space (galactic center)	Visible light	5×10^{-4}	5×10^{15}
Earth surface	Visible light	3×10^{-3}	3×10^{16}
	Ultraviolet light	1.6×10^{-4}	2×10^{15}
Sun surface	Visible light	625	6×10^{21}
	Ultraviolet light	60	3×10^{20}
Neutron star surface	Cosmic radiation	4×10^{18}	10^{26}
White dwarf surface	Radiation from interior	2×10^7	5×10^{26}

Some flows of energy in representative environments through-out the Universe. These flows are expressed both as total amounts of energy and number of individual energy packets. Locations are listed in order of increasing number of packets. Note the wide variations in the flow.

flows. A summary of the free-energy flows in various environments is given in Table 7-1.

STORED ENERGY FOR LIFE

As an alternative to using a steady flow of free energy which enters the environment, life can function by using energy that has been stored in an environment over a long period of time. The choice between these types of energy has become familiar to us recently through the discussion about using nonrenewable (stored) energy sources such as petroleum versus renewable sources (flows) such as solar energy for human activity. In some cases the amounts of energy that can be extracted from stored pools are comparable to what is available from some of the energy flows discussed above. However, extracting energy from storage involves more effort on the part of a living thing than tapping an energy flow, since it becomes necessary to move around from place to place when the energy stored in one region becomes depleted. This problem is not insoluble, and we see no reason why stored energy should not be the source for a variety of life forms. But its usefulness depends on the density of stored energy, the rate at which this energy is replenished, and the speed at which a living thing can move from place to place in order to find energy that it can feed upon. Also, it is important how any form of energy couples to the vehicle by which order is stored. If many steps are necessary to transform the free energy into stored order, the life form will have to be quite complicated in its metabolism, and probably will require more energy than a simpler living thing to produce the same amount of order. We will discuss each of these aspects for some different types of stored energy.

A temperature difference may be considered a form of stored energy. Wherever such a temperature difference exists between two places, free energy can be obtained by taking in heat at the higher temperature and exhausting it at the lower one. If the two places are maintained at a fixed temperature difference, then the maximum amount of free energy per cycle that can be obtained in this way is just proportional to the temperature difference. Otherwise, as energy is extracted, the two temperatures come close together and the process cannot continue indefinitely.

Examples of such temperature differences exist inside the Earth and within the oceans. The temperature differences in the Earth are maintained by the radioactive decay of uranium and thorium in the interior. The ocean surfaces are kept warmer than the bottom by the sunlight absorbed near the top. The temperature differences produced this way on the Earth's surface are not great enough to make it useful as a direct source of free energy for life. In other parts of the Universe, much larger temperature differences can occur and perhaps be used for life processes.

In order for this to be possible, the living thing must move from a place where the temperature is high to one where it is lower, and back, over and over again. Alternatively, it might extend from the cold to the hot area and could transfer the heat internally. Therefore, a relevant quantity is the rate at which temperature changes with location. A few examples of this quantity are given in Table 7-2. This table shows that in a variety of environments the temperature changes by a few degrees per kilometer. The energy that can be extracted from a temperature difference depends on the process used. Imagine a living thing that can shuttle back and forth between regions of different temperature at a speed of ten kilometers per hour. Suppose that the creature has an internal supply of gas whose temperature can increase or decrease through contact with the outside environment. This process can extract energy from the temperature difference through the expansion and contraction of the gas that accompanies its change in temperature. Such a creature, in one of the environments shown in Table 7-2, might be able to extract as much as one-fortieth of a calorie per second, per gram of its internal gas. If it is to obtain as much

Table. 7-2

TEMPERATURE CHANGE WITH DEPTH

Environment	Rate of Change of Temperature (degrees centigrade per kilometer)
Earth crust	10
Earth mantle	1
Sun interior	5
Jupiter	2
White dwarf interior	100?

How the temperature varies with depth in different environments. In each case the temperature increases going inward.

power per unit of its total mass in this way as typical large Earth creatures can by absorbing sunlight or by eating plants, then it would require an amount of internal gas equal to a few percent of its total mass, which is not an unreasonable requirement. This suggests that for environments where no richer source of energy is available, creatures which extract energy by commuting between regions at different temperature may be the dominant form of life.

It is also possible that in some environments there are stable situations in which nearby areas differ in temperature by greater amounts than in the environments listed in Table 7-2. In such cases it would be easier for living things to extract energy. We will use the name thermophages (heat eaters) to describe life forms that extract free energy from temperature differences. We will encounter them again when we tour the planet Jupiter.

The type of stored energy we are most familiar with is chemical energy. We take it in daily, usually at the rate of three meals a day, and use it to keep our bodies functioning and in order. The amount of stored chemical energy varies a lot from place to place on the Earth's surface, and not all of it will serve as food for us. In such microenvironments as a pool of oil or a drop of milk, the stored energy density is very high, while in a snowdrift it is quite low. Much of the stored energy on the Earth's surface is chemical energy in the form of hydrocarbons such as oil, pure carbon like coal, and living material such as vegetation. There is actually much more stored nuclear energy in the form of hydrogen in the ocean's water, but it is not yet accessible to life. Eventually, if human civilization runs on nuclear-fusion reactors, we will have tapped into this new source of stored energy.

As stated before, the use of stored energy depends on a living thing being able to move from place to place to feed on it. Generally speaking, the quantity which can be compared to an energy flow is the product of the density of some type of stored energy, multiplied by the speed at which a living thing can move through the stored energy, using it as it goes. Large animals, such as cows, would do much better by grazing on grass than by absorbing sunlight, even if a cow were capable of photosynthesis. On the other hand, once the grass is eaten, it takes a while to grow again, whereas the Sun shines anew each day, which is one reason why life is easier for plants than for animals.

While individual living things on Earth can feed on stored energy, it does not seem possible for the whole biosphere in its present size to do so for any long period of time. The total amount of chemical energy stored near the Earth's surface in hydrocarbons corresponds to only a few thousand years of energy use at the rate at which the biosphere now uses energy. Unless some parts of the biosphere learn to live off the huge supplies of stored nuclear energy, there is no plausible alternative to sunlight that is presently available on Earth for the biosphere.

Nuclear energy is actually the most plentiful source of stored energy in much of the Universe. It can become available when certain types of hydrogen nuclei are fused into heavier elements. Most stars make use of this process as their primary source of energy. Is it possible that a life form might find a way to tap the potentiality of nuclear fusion to fuel its life processes? As we have stated, human beings are on the verge of using fusion energy industrially. We have not considered using it biologically, because this type of energy is not well matched to the nucleic-acid mechanism for storing order. It is imaginable that life forms using other mechanisms for storing order might have reasons to develop methods to accelerate nuclear fusion. Several such methods have been suggested in order to meet human industrial energy needs, including high pressures, high temperatures, and catalysis of fusion by subatomic particles. It is uncertain if any such methods, or others not yet known, would really work to make fusion possible on a substellar scale. But in view of the immense quantities of hydrogen present in diverse parts of the Universe, and of the relatively large amounts of energy that the fusion of two hydrogen atoms makes available, it is plausible that somewhere living things have evolved which can use this most primitive of all elements, deriving their nourishment from the same source that makes the Sun shine. Chapter 12 contains one suggestion about such creatures in connection with possible life in the Sun.

Perhaps the most useful form of stored energy is one that is constantly replenished through the action of an energy flow. If this happens, the stored energy is really just a convenient intermediary between the energy flow and its ultimate consumer, just as a plant is an intermediary between the Sun and a hungry herbivore. But the energy does not need to be stored through the activity of some living thing. There are simple physical processes that can lead to energy storage on a cosmic scale. One way in which this can happen is through the effect of radiation on atoms and molecules. Any given type of molecule or atom can exist in many different internal configurations, each with different energy or different amounts of rotation. Ordinarily, in familiar conditions most atoms and molecules are found in the configuration of lowest energy, called the ground state. However, some are always present in configurations of higher energy called excited states, especially in the case of molecules, where the energy differences are comparable to the energy of thermal motion of the molecules. It is possible through the absorption of radiation by the atoms for a large number of the atoms in some region to be kept in an excited state for an extended period of time. This state represents a deviation from equilibrium and extra free energy is stored in the excited atoms.

At most temperatures the energy per atom of such excitations is much greater than the heat energy associated with the atoms, which suggests that it should be easy to use this energy to sustain life. However, this would depend on the ability of a living thing to obtain a sufficient number of excited atoms to maintain itself. If only a small number of atoms are in the excited state, the trouble of finding the excited ones may outweigh the gains when they are found. A ten-thousand-dollar bill is nice when you get one, but it is better to rely on ones and fives for everyday needs. One plausible environment where a sufficient supply of excited atoms might exist for a scavenger would be in the neighborhood of certain stars or even in their outer layers. High gas densities exist in these areas, as well as a high flux of radiation.

We can summarize the energy prospects for life in the Universe as follows: In almost all the regions where there is sufficient matter to make life imaginable, there is enough free energy in one form or another to act as an ordering agent for life processes. Measured in terms of the number of energy packets available per atom, the energy flows and the energy stores available in diverse environments are comparable to those that have allowed life to flourish on Earth's surface. Only in the spaces between galaxies, where the starlight is thinned out by distance and the microwave background reigns supreme, is there insufficient energy. But in these spaces also, the densities of matter are tiny, and even plentiful sources of energy would have little to work with. Within the galaxies, and especially near and inside the dense accumulations of matter we call stars, free energy both flows and is stored in amounts sufficient for life to thrive, providing that other constraints can be met.

MECHANISMS FOR STORING ORDER

A supply of free energy is not sufficient as a basis for life. There must be an appropriate form of matter to interact with it. There must also be a suitable coupling between the type of energy and the matter that absorbs it. The result of the interaction must not be just to heat the matter up but rather induce order in it.

How can order be produced? A number of stages are involved. First, there must be states away from equilibrium that are accessible to the matter under the influence of the energy flow. These states may contain some degree of order, but it is their potentiality that is most important. They must contain the possibility of becoming progressively more orderly through the further action of the energy flow. According to our discussion of order, this can happen if some property of the system can occur in many different forms, such as the many chemical combinations of the atoms C, N, O, H. Order is achieved through the selection and preservation of a small number of these forms. Eventually, a degree of complexity is reached which we would associate with living matter.

A key point in this scheme is that the first nonequilibrium situations which are produced, and the later ordered situations, should exist long enough that the progression toward greater order can continue. For this to happen, the inflow of free energy that produces the order must be sufficient to overcome the reactions within the system that tend to restore it to equilibrium. While it is important that the components of a system react quickly enough for evolution towards order to occur, if they react too quickly, order cannot be produced either. This is because a too rapid reaction rate reduces the ability of a flow of free energy to produce order. In the poker game of Chapter 3, if the money goes into the kitty too rapidly, the external source of money will not be able to keep the game going.

It is also important that random environmental changes, which can disrupt order, not be too strong. If they are, the situation is analogous to a child's effort to build a sand castle in the face of an increasing tide. As the waves continue to come in, it becomes impossible to repair the castle or add height to it, which would represent the maintenance or increase of order. In a more favorable situation for life, random environmental forces will not be strong enough to disturb the base of order, which will be affected only by the specific flow of energy that creates it. A sand castle that is built far from the surf can usually resist the wind from the sea, the analogue of the weak environmental forces.

Let us apply this generalized account to the steps that may actually have occurred in the origin of life on Earth. The available energy flow was primarily visible and ultraviolet light rays from the Sun. The initial system was the prebiotic atmosphere of Earth, a mixture that may have included water, nitrogen, hydrogen, and methane or carbon dioxide. It was very tidy, in that it contained few components, but it also had little order, as it was close to equilibrium. The energy flow through this mixture converted it to a more ordered, but less tidy, array of small organic molecules containing bonds from carbon to hydrogen, nitrogen, oxygen, and other carbon atoms. This was the prebiotic soup.

The relatively strong chemical bonds that were formed made the prebiotic soup fairly stable against reverting to the starting materials through collision processes. On the other hand, it could use the continuing flow of solar energy, as well as the stored chemical energy present in its own bonds, to proceed to an even more ordered state. There are at least two different paths by which this could have occurred.

If the formation of more bonds of carbon to nitrogen, oxygen, and carbon had continued indiscriminately, the product would have been a random polymer containing a large number of atoms forming a network of bonds extending spatially in all directions. This result often occurs in unsuccessful reactions in organic chemistry. The products are tars and resins, which coat the bottom of the reaction flask. Instead of producing more order, the flow of free energy has been stored as chemical energy in the resinous polymers. The low reactivity of these polymers and their insolubility removes them from any further ordering transformations. They represent a dead end to chemical evolution.

Luckily, another path was followed at least once on the prebiotic Earth. The initial mixture of small molecules interacted with the energy flow in another way. A select subset of the original mixture of chemicals was preferentially formed and concentrated. This mixture was further refined until it gave rise to a very few well-selected larger molecules, which possessed significant capacity for catalysis and the storage of order. Life on Earth had begun.

It is easy to devise scenarios in which one or more vital ingredients in the recipe are missing, and the interplay of energy and a mixture of chemicals does not lead to increasing order and ultimately to life. For example, if a planet were made entirely of hydrogen and helium, little in the way of interesting chemistry could take place. (Although, as we will see later, physical processes capable of ordering matter could occur.) Suitable radiation could break hydrogen molecules apart into atoms, but they would have little to do except recombine. On the other hand, if even small amounts of other elements are present, the hydrogen atoms could react with them and perhaps produce an interesting biochemistry. Life need not use the most common materials present in a particular environment. One reason is that the principal materials may have the wrong properties to become ordered under the prevailing environmental conditions.

An illustration of this point exists on our own planet. Sand and rock are made of chemicals called silicates. Under the conditions on the Earth's surface, silicates are too stable chemically for much to have happened to them. For this reason and others, sand life has not evolved on the surface of the Earth.

Other impediments to the development of order may exist. Many planets and satellites have a solid surface but no liquid or atmosphere. Light may produce excited species by interactions with the chemicals on the surface, but frozen in place in the solid state as they are, they cannot interact with each other effectively to produce more complex molecules. Similarly, mobile excited molecules may be formed in intergalactic space, but they cannot locate each other fast enough to react chemically.

The reasons for failure in the above cases are obvious. It would be ludicrous for us to conclude from them, however, that chemical life can succeed only if the Earth recipe is followed in every detail. It is quite possible that a different subset of the many compounds of carbon, hydrogen, oxygen, and nitrogen than that found in our own life could serve as the basis of chemical order, even under conditions similar to those on Earth. For example, nothing we know at present rules out the possibility of a world in which amino acids occur only in the mirror-image forms of those found in Earthlife. Or in another biosphere a completely different set of amino acids might predominate. Even within the confines of carbon chemistry, there are myriads of other solutions to the problem of storing order chemically. We do not know why Earth's biosphere has chosen the set of chemical compounds which it contains. Conceivably, it is because these compounds have some subtle relationship among their properties that allows them to fit into a self-perpetuating chemical "community," and no other collection of compounds could do this. More likely, there was an interplay of chance, environmental factors, and some specific properties of Earth biochemicals that led to the enhanced production of these substances. But such a result cannot be universal.

The exact set of chemicals present on Earth is thus not essential to life. The same can also be said about the temperature, the liquid medium, and the particular elements that are used. It is true that severe changes in these variables would create an environment unsuitable to Earthlife. But other combinations leading to other life forms could arise as an alternative.

Changes in temperature alter the reaction rates between different molecules in different ways. As a result, a collection of molecular reactions which form a stable cycle at one temperature, so that the rate of production of each chemical balances its rate of destruction, will not do so at another temperature. Conversely, a temperature which is unsuitable for the existence of cycles among one set of chemicals might still be suitable for sustaining a different cyclic set of reactions based on other chemicals.

The temperature of the environment affects not only the rate of reactions, but also the stability of various molecules and combinations of molecules. As the temperature increases, the violence of molecular collisions also increases. Only those molecules which are held together by bonds that are substantially more energetic than the average energy of motion of the molecules can be expected to be stable at any given temperature. Atoms and molecules are bound together in various ways, all manifestations of the fundamental electrical force that acts between the charges of which both are composed. The resistance of various bonds to breakage depends directly on the enviromental conditions and the specific type of bond involved. Some of the molecular bonds, known as covalent and ionic bonds, are very strong and require as much energy to break them up as to pull apart their individual atoms. The hydrogen bonds that hold DNA together break under some circumstances at body temperature, and seldom survive temperatures much higher than 100°C. The stronger covalent and ionic bonds could serve the same purpose for life at higher temperatures that hydrogen bonds do at Earth temperatures.

Similarly, many substances in Earthlife would be too stable to be utilized in life at lower temperatures. There are many chemical bonds, however, which are too unstable to exist at Earth temperatures, but would be suitable as a basis for life at lower temperatures. There is nothing about the temperature range found on Earth that makes it uniquely fitting for complex biochemical processes. This topic will be treated in greater detail in Chapter 8, when we consider specific alternatives for chemical life.

The supposedly unique properties of water as the medium for life, and of carbon compounds as the basis for sustaining a chemistry sufficiently complex for life, have been celebrated to such an extent that they comprise a large portion of the literature on extraterrestrial life. These arguments will be given careful consideration in our chapter on chemical life. At this point, however, we will just say that we disagree with them and turn to an even more general point about vehicles for order.

Any kind of vehicle for order that depends on the use of specific types of atoms or combinations of atoms can operate only if they are present in sufficient quantities in some environment. The total number of atoms, as well as the relative numbers of different types, varies tremendously from place to place in the Universe. Some figures are given in Table 7-3 for a few selected environments, as well as for the average composition of Earthlife. An interesting point shown in the table is that the atomic composition of Earthlife is more similar to the average composition of the whole Universe than to that of the Earth's surface. This could suggest an extraterrestrial origin for Earthlife, perhaps on a planet like Jupiter, which is much closer to the "universal" composition. However, it is more likely that the atoms composing Earthlife are those that function best in living matter under the other conditions of Earth's environment, and so have been naturally selected as its components.

One general feature of the data in Table 7-3 is that heavy

Table 7-3

COMMON ELEMENTS IN VARIOUS ENVIRONMENTS

Stars and Interstellar Gas	Earth's Surface	Dry Biomass
(1) Hydrogen	Oxygen	Oxygen
(2) Helium	Silicon	Carbon
(3) Neon	Aluminum	Hydrogen
(4) Oxygen	Iron	Nitrogen
(5) Nitrogen	Calcium	Phosphorus
(6) Carbon	Magnesium	Sulfur

The most common elements found in various environments. They are listed in order of their occurrence, starting with the most common.

elements are exceedingly rare in most of the Universe. Six types of atoms—hydrogen, helium, oxygen, nitrogen, carbon, and neon, all among the ten lightest—make up 99.9 percent of the atoms in the Universe. The only large regions we know of that contain significant fractions of heavier elements are the interiors of old stars, where these elements are being produced but where the temperatures are too high for life based on chemistry; and the terrestrial planets, in which the lighter elements have been selectively driven away, leaving higher concentrations of heavier elements. This process is similar to the evaporation of water from a salt solution, which leaves the salt behind. Therefore, even if life can use heavy elements as a vehicle for order, we do not know of many places where they are available in quantity, and we might expect heavy-element life to be rare on that account.

However, on Earth the processes of selective concentration have produced a crust rich in silicon, aluminum and iron. One can wonder whether such processes might produce environments especially rich in other heavier elements. Probably the right combination of environmental conditions could relatively enhance any element almost as much as any other. Some of the elements that are as cosmically abundant as iron or aluminum are magnesium, calcium, sulfur, and nickel. The first two are fairly common on the Earth's surface, the last two somewhat rare. Both sulfur and nickel are reasonably active chemically, and it would not be surprising if they were as common on some other planets as iron, silicon, or aluminum on Earth. Actually, we know of microenvironments, some of the asteroids that orbit the Sun between Mars and Jupiter, in which the predominant materials are heavy elements such as nickel and iron. Had the temperature history of the Earth's surface been somewhat different, it is possible that more of the volatile elements like hydrogen, oxygen, and carbon would have been driven off, leaving almost a pure residue of heavy elements. For these reasons it is worthwhile to consider life forms based on heavy elements, since environments where these elements predominate may not be as uncommon as their overall abundance would suggest. Indeed, one such environment is literally under our feet—the interior of the Earth. The center of our planet, according to geologists, is almost completely a liquid of heavy elements such as iron. Various physical processes may occur in such heavy-element liquids which might form a basis for life.

OTHER MECHANISMS FOR STORING ORDER

Mechanisms for storing order in matter exist which do not depend on the occurrence of complex mixtures of reacting chemicals. For example, order can be created and stored in some of the variable properties of individual molecules and atoms.

We discussed earlier that atoms can exist in a lowest energy state or in excited states. The excited states were considered as a possible energy supply or food for life. In another context they could be used to store order, as part of the material basis of a life form. At any temperature there is an equilibrium pattern with most atoms in the ground state and a few in excited states. This situation can be regarded as having no order. Any deviation from it, say one with many atoms in excited states, represents some amount of order. Alternatively, there can be many different ways in which an atom can have the same energy but have different values of other qualities, such as its rate of rotation. In the disordered situation, these different ways of having the same energy tend to occur equally often. If some of them are preferred, there is again some order to the system. A specific arrangement of atoms, for example one in which the internal energy state is correlated with the position of the atom along a chain, is a type of order very analogous to the spatial order of nucleic acids, where the type of base is correlated with its position in the chain. But the distribution of atoms over internal energy states does not have to be spatially correlated to represent order, any more than the preferential occurrence of certain organic molecules on Earth requires a particular spatial distribution to represent order. Just having a distribution of energy states that are different from the equilibrium distribution is orderly.

In general, if left alone, extra atoms will not remain in excited states for very long and tend to revert to the lowest energy state by emitting radiation. However, in the presence of incoming radiation of the right wavelength, or of other forms of free energy, atoms or molecules can be maintained in excited states indefinitely. This actually happens in some stages of the action of a laser. The method of maintaining order by a flow of free energy through a system is very similar to what Earthlife does to maintain its chemical order. The intense energy flows of various types in some parts of the Universe may produce this "laser" type of order. Some evidence exists for natural lasers in certain interstellar regions, where it is thought that microwave radiation acts on molecules to keep a large fraction of them in excited states.

This way of ordering the energy levels of atoms or molecules, especially through a flow of radiation, does not depend significantly on having a high density of atoms present. Indeed, such a high density may hinder the ordering process, because the outer atoms may shield the inner ones. This type of order may therefore be more appropriate for life arising in low-density regions, such as interstellar gas clouds or the outer layers of some stars. A life form based on the interplay of radiation and matter at low densities is interesting enough to be considered at length and will be covered in Chapter 12.

Other life forms not based on chemistry could develop in very dense forms of matter which exist in parts of the Universe. Since atoms are not the usual form at such densities, order must involve something other than the types of atoms or their combinations. Because there are stable forms of matter at high densities, it is among them that order must develop if life is to exist. In some kinds of dense matter, the electrons become separated from their nuclei and form a kind of gas in which the nuclei are suspended. This type of matter is found in certain aged stars known as white dwarfs. There are other, still denser forms of matter, most of which are composed of neutrons that are close enough to each other to be in contact. These conditions are found in objects called neutron stars.

The general conclusion that can be drawn from these discussions is that possibilities for storing order exist in almost any environment we know of in which matter is present. This qualitative point must be made more quantitative, and to do that we must estimate the amount of order that would be relevant to a process identifiable as life. Individual living things on Earth may contain between 10^{10} and 10^{32} atoms, while the Earth's biosphere contains some 10^{47} atoms. A rough approximation to the amount of order in a highly ordered system of atoms is that it is proportional to the number of ordered units. For the biosphere this is approximately the number of small organic molecules, or about 10^{41} units. For an individual living thing, the amount that should be used is probably the number of bases on the nucleic-acid strand, which is much smaller than the number of atoms and varies in bacteria from several hundred thousand through several million. If we take the smallest estimate, we could say that an object needs at least several hundred thousand ordered units in order to qualify as living. This may possibly be an overestimate, as some viruses, which are "almost" living things, have less than ten thousand bases in their nucleic acid, and viroids, as we have seen, can have just a few hundred. On the other hand, if we are correct in thinking that life can arise only in the form of a biosphere, it is possible to conclude that a much larger number of ordered units is necessary for life to exist.

There are several reasons why a substantial number of ordered units should be necessary for life. One is that life containing a very small number of units will be unstable against the chance fluctuations which are unavoidable in any physical system. For example, if a life process involves some chemical reaction which usually proceeds in one direction, there is some chance that the reaction may in one instance go in the opposite direction. If the living thing contained only a few atoms, this instance could be a serious disturbance to its life process; while if it contained many atoms, a single "wrong" reaction would not likely have a serious effect.

A second reason for requiring many ordered units in a living thing is so that it can perform a variety of functions. One reason that most biologists do not think of viruses as alive is because viruses can do so few things that they require the machinery of another cell to function. It does seem a reasonable requirement for a living thing to be able to do "enough" on its own to qualitatively distinguish it from non-living processes, but it is hard to make this criterion a quantitative one. In Earthlife the minimum number of units is determined mainly by the requirement of coding a number of different proteins, each of which contains hundreds of amino acids and so requires many hundreds of bases to determine its structure. However, there is no obvious reason why simpler molecules than proteins could not function as catalysts in chemical life in other environmental conditions, in which case the requirements for coding them would be reduced. It is even possible that the functions of order storage and catalysis might reside in the same molecule, rather than in two different types. Accordingly, even within the framework of chemical life, living things might occur whose genetic information consists of no more than a few hundred units, and which nevertheless perform enough functions to be interesting.

The question of how small a biosphere can be is much harder to answer, because here we have only a single example from which to draw conclusions. It may be that what is important is not the absolute size of a biosphere, but rather its size relative to the environment that is not a part of it. While the biosphere is able to maintain its own order by using free energy, if it is in contact with a large environment there will be a tendency for the order of the biosphere to become degraded through its interaction with the matter in the environment. For example, if Earth's biosphere were limited to a single pond of water, it is possible that life would not persist longer than the first dry summer. One way to escape such a fate is for the biosphere to encompass most or all of the interacting material in some region. If this can happen, then the biosphere can proceed in its own way, without contamination by outside influences which do not take part in its ordering cycles. This is the path that has been followed by Earth's biosphere with a certain degree of success.

Even if the biosphere remains in contact with some matter which does not take part in its functions, it may still be relatively immune to disordering if it is large enough that only a small part of it is vulnerable. This is the principle which leads organisms to be arranged so that most of their vital organs are inside a relatively impermeable skin. However, there may be a conflict between this approach and the biosphere's need to exchange free energy with the environment. A biosphere that is too successful in walling itself off soon dies from a lack of free energy. But it is not difficult in principle to devise a "skin" for a biosphere which would allow some types of energy in and out yet would be impervious to those forms of matter and energy that would destroy the biosphere as they enter or leave. Lewis Thomas has argued convincingly that Earth's atmosphere acts as just such a skin, allowing in visible light but not ultraviolet or ionizing particles and radiation. This type of solution is possible only if the biosphere is large enough, or the skin would have to be too large a fraction of the total size of the biosphere to provide protection. The precise absolute numbers depend on the specific environmental conditions, and we will discuss some examples of plausible biosphere sizes in Chapter 12.

IS THERE ENOUGH TIME FOR LIFE?

When a sufficient energy source is interacting with a suitable medium for the accumulation of order, two of the crucial criteria for life have been met. The remaining need is for sufficient time to bring the process to fruition. How long it may take for a given type of order to develop in a certain environment is the most critical question that determines the scope of life in the Universe.

We can think of two types of mismatch between the time needed and the time available. The rate of a process, such as the reactions between molecules widely scattered in outer space, may be so slow that the entire lifetime of the Universe to date has been insufficient for appreciable order to develop. In another case a drastic change in an environment in which order has been accumulating at a reasonable rate may destabilize or destroy the base upon which order has been built. For example, the entire contents of a planet which contains developing molecular life may be roasted and converted to unordered atoms when the parent star becomes a supernova.

Less extreme environmental shifts can also have adverse effects on a life form, but the outcome is not always so obvious. An environment in which important characteristics, such as temperature, change very quickly may be inhospitable to life, because living things cannot adjust their behavior rapidly enough to survive the changes, even if the average conditions are hospitable. Moon's surface has an average temperature similar to Earth's, but the change in temperature from day to night on Moon is so extreme that it would be hard for an organism to survive under both conditions, although perhaps not impossible. The significance of the requirement that conditions not undergo drastic change too rapidly should not be overemphasized. On one planet's surface, the energy flow changes periodically from a few calories per second per square meter to about zero within a space of a few minutes, thus depriving a large proportion of the living things of their energy supply. This phenomenon, known to humans as the sunset, has not stopped plants from evolving on Earth. Similar things could happen in other environments without making them too unstable for life to develop there.

In considering whether there is enough time for life to develop in a given environment, we therefore must consider the stability of various localities and the time needed for life to develop. Of all the various requirements for life discussed, the time needed for evolution is the one whose significance is the most difficult to evaluate quantitatively. We do not know how long it took for the first forms of life to develop on Earth. Estimates have ranged from seven days in the Book of Genesis up to a billion years or more. Even if we knew this, it would not determine how long life, which might perhaps use different aspects of physics or chemistry for its workings, would take to develop in other environments. We shall see that the rates of the fundamental physical processes upon which life might be based vary tremendously from process to process. It seems likely that the rate of evolution must be somewhat correlated with the rate of these underlying processes. Therefore, it appears a plausible guess that different life forms have evolved over immensely different periods of time, depending on their environment and their basis.

One consequence is that some environments, which last relatively short times on an astronomical time scale, may nevertheless be homes for life of a type that has developed very rapidly. For example, some very massive stars run through their whole cycle of energy-generating processes in a hundred million years or so, a small fraction of the time needed for higher organisms to evolve on Earth. Nevertheless, the planets of such stars, or even the stars themselves, might be inhabited by living things functioning by processes whose rates are much quicker than the chemical processes of Earthlife, and which therefore could evolve in the short time available. Again, there is no guarantee that a perfect match will occur between the time over which some environment is relatively stable and the time needed for a life form viable in that environment to evolve, but enough possibilities exist that we cannot rule out short-lived environments as homes for life.

The time scale for the development of life depends on at least two critical factors: the rate at which energy becomes available and the reaction rates of the processes involved in building up order. It is the second factor which can vary most from case to case and which is the hardest to estimate. Furthermore, when many different reactions are involved in the life process, it is the slowest reactions which will act as the bottlenecks governing the overall rate of development. Therefore, we must identify the possible bottlenecks that can limit the rate of development of various forms of life from primeval conditions.

For chemical life an important bottleneck is that the atoms or molecules involved in the chemical reactions must be free to move around and find their partners. The mobility of the atoms tends to be high when they are in liquid solutions or in gases, but quite low when in a solid. If we compare reactions in solution with those in a gas or a solid under Earth conditions, the reactions in solution are often much more rapid, for two reasons. One is that the reacting atoms can move about freely in solution and find their reaction partner. In solids, on the other hand, the atoms tend to be fixed in place so that each atom can only react with a few nearby, and the opportunity to find a partner is much less. The situation in a solid is somewhat similar to a bachelor trying to find a mate while seated in a movie theater, as opposed to the situation in a solution, which is more like visiting many singles bars.

In comparison with a gas, atoms in a liquid have somewhat less mobility, but there are higher concentrations in the liquid because the density of liquids can be much greater than that of gases, at least at Earth pressures. This would not be true for the highly compressed gas found in the lower atmosphere of planets like Jupiter, whose atmospheric density is one thousand times greater than that of the Earth's atmosphere at sea level. In such an atmosphere, the molecular concentration could be as great as in most liquids. However, there is another important effect that some liquids, such as water, have on reaction rates. Molecules dissolved in certain liquids can break up into positive and negative ions. Reactions involving ions of opposite charge can often proceed much more rapidly than those involving neutral molecules. Gas-phase reactions, which generally involve neutral molecules, may go more slowly than those involving ions in liquids, even at similar density. On the other hand, reactions between ions whose charge has the same sign may proceed more slowly than those of neutral molecules. The overall effect of molecular dissociation into ions on a linked set of chemical reactions is complex, and it is not easy to predict whether it speeds up the set or slows it down.

These arguments suggest that while the presence of

liquids as a medium for chemical reactions may be convenient, it does not seem absolutely essential. A dense gas, especially at the high temperatures which enhance reaction rates, could be an equally convenient medium for chemical reactions, and one which is much more common in the Universe than liquids.

The rate of chemical reactions in any specific medium depends on the temperature as well as on the atoms that are reacting. Usually the rate for a reaction involving specific atoms increases with temperature, and a significant problem for low-temperature life could be that these rates become too small. However, this effect can be somewhat compensated by changing the atoms involved. There are reactions that are extremely rapid at Earth temperatures, such as those involving individual atoms rather than molecules. These reactions, at low temperatures, can still have rates comparable to those of many biochemical reactions at Earth temperatures. Furthermore, if a reaction goes too slowly as a one-step process, it sometimes can go more quickly through a series of intermediate steps, each of which is a fast reaction. This is how some catalysts work to speed up reactions. Also, the time that it takes for a given atom to react is inversely proportional to the concentrations of the atoms it reacts with, as well as to the reaction rate. There is also a tendency for these to compensate for each other. If a reaction rate is small, the number of atoms reacting can remain high for longer than if the reaction rate is large, providing that most of the atoms are part of the reacting pool.

As a result of all these influences, we might expect that at low temperatures the time scale for chemical life would be longer than at Earth temperature, but not according to any simple rule. Molecules which can react relatively quickly at low temperatures are likely to be substituted for those that occur in Earthlife. These would include "free radicals," molecules which react so quickly at Earth temperatures and densities that they usually can exist only transiently. However, at low temperatures they may exist for much longer times. Such free radicals have in fact been observed in interstellar space under conditions of low temperature and density. The very observation of free radicals and other molecules under interstellar conditions indicates that chemistry can take place in very different contexts than those familiar on Earth. Harold Urey has remarked that ". . . the apparently unusual conditions under which chemistry occurs in space are really the ordinary conditions for chemical phenomena in nature."

The decisive feature for chemical life in much of the Universe is not the temperature but the low density. In most interstellar and intergalactic space, there is less than one atom in each cubic centimeter of space. This density is comparable to having a single bacterium, and nothing else, in a football stadium. At this low density, encounters between atoms, which are necessary for chemical reactions, are very rare. The average atom meets another atom only about once in a thousand years or so, depending on its state of ionization and the ambient temperature. Under these conditions it would take something like a million years for an object the size of a bacterium to interact with enough atoms to double in size, not a promising time scale for a chemical life form.

At the much higher densities in some dense clouds, the time scales for atomic encounters can become much less, perhaps as little as a year. The time that it takes an object to double its size through atomic encounters, if every atom pair sticks together, varies inversely as the density of the surrounding medium. For densities of 10^{12} atoms per cubic centimeter, the time would be a few minutes, comparable to the actual doubling time of Earth bacteria. This density is higher than that of any known stable interstellar cloud. It is roughly intermediate between the densities in typical interstellar space and those in condensed solid bodies such as Earth. However, even at higher densities the problem is that most of the atoms you will meet are hydrogen and helium atoms, which are not suitable substances on which to base a complex chemistry. It is hard to avoid the conclusion that the interstellar environments with which we are familiar were not suitable for the development of chemical life over the history of the Universe until now.

The question is whether there are yet unidentified interstellar regions, say of much higher density, where the time scales for atomic encounter could be short enough for chemical life to develop. Certainly there must be local regions in which the densities are greater, since we know that stars originate in the interstellar gas, and in stars the average density of matter is similar to that on Earth. The problem is that such high-density regions tend to collapse rapidly into stars, presumably carrying any interstellar life along with them. There is an approximate relation between the maximum density and size that a gas cloud can have without collapsing under its own weight. Some examples of this relation are given in Table 7-4. Actually, if the cloud is rotating, it can be somewhat larger for a given density without collapsing, but not very much so. It can be seen that gas clouds which are as large as the solar system can reach respectable densities without collapsing. Conceivably, in such clouds, which would be to small for us to detect, the time needed for growth by adding on molecules could be short enough for chemical life to develop.

There are other environments containing extended gaseous regions whose densities and temperatures are in the general range that would allow a rate of encounters sufficient for chemical life. These include the atmospheres of giant planets, such as Jupiter, and the outer atmospheres of certain cool stars, known as red supergiants. In these environments it is not the rarity of atomic meetings that is the bottleneck chemical life must get through; instead, more subtle problems restrict the rate at which such life can evolve.

The processes of chemical life are a series of connected reactions, each of which influences the others. In order for any one of the catalytic cycles involved in life to work, precise amounts of various chemical substances must be present in the right place at the right time. A catalyst for reaction X

Table 7-4

DENSITY VERSUS SIZE OF GAS CLOUDS

Density (atoms per cubic centimeter)	Maximum Size (centimeters)	Maximum Mass	Average Time Between Collisions of Atoms (seconds)
1	10^{20}	10^4 times mass of Sun	10^{11}
10^4	10^{18}	100 times mass of Sun	10^7
10^8	10^{16}	mass of Sun	1,000
10^{12}	10^{14}	10 times mass of Jupiter	.1
10^{19}	10^{11}	mass of Earth	10^{-8}

The maximum size that a gaseous cloud can have without collapsing under its own weight. Some properties of these clouds are also listed. The numbers are calculated for a cloud temperature of $30°C$, and depend somewhat on the actual temperature.

is of little use if it is produced long after the molecules that should react have been depleted by other processes which compete with X. The need for such precise balances in a series of reactions is likely to extend the amount of time that a whole cycle takes far beyond that of the individual reactions. This is especially true when a form of life is first developing, and the different reactions have not become synchronized. Imagine a relay race in which the runners on one team start independently, but the baton must still be passed in a certain area in order to avoid a foul. The runners will have to adjust their speeds, and the race will take much longer than what one might expect by averaging the speed of each runner. Only if the runners have some way of signaling their intentions in advance is it possible to arrange the race so that each runner arrives at the required spot precisely with the next runner to whom he must pass the baton. The latter situation is analogous to what happens in developed life, but it could not have happened when life was first beginning. This type of rate synchronization is one of the things which must develop in the pre-Darwinian evolution that occurs before the biosphere comes alive. Perhaps it should be thought of as another kind of order, one which arranges things in time rather than in space. Although we have described this synchronization problem for the case of chemical life, it exists for any type of life that involves integrated cycles of processes, that is, probably for any type of life at all.

It would be helpful to have guidance from experiments about the conditions under which an unordered set of chemicals, subjected to an energy flow, develops a given amount of chemical order. Most prebiotic experiments to date, unfortunately, were run for only a fixed amount of time. At the end of this period, an analysis was conducted for the presence of certain key compounds, such as amino acids, but no total inventory of products was attempted. It would be very exciting if conditions were found which, under an uninterrupted energy flow, demonstrated a measure of chemical evolution. For example, the presence of certain chemicals might catalyze the formation of others, so that a restricted number of components dominated the mixture. Even if the chemicals differed from those prominent in Earthlife, a study of the cycles involved might give us some idea of the time needed for the pre-Darwinian evolution of chemical life.

In the absence of this information, we can try to estimate this time theoretically, but it is a difficult mathematical problem. It seems plausible that for reactions which proceed independently, the amount of time that it takes to produce a specific set of products increases very rapidly as the number of products involved increases. This increase is much more pronounced than for synchronized reactions, where the time increases about proportionately to the number of steps. Therefore, the main time requirement is the amount that is necessary to achieve some synchronization, after which additional steps could be added more rapidly.

Some information about the time involved may be determined by studying what happens in Earth's atmosphere when a new man-made chemical is added inadvertently. Several such cases have occurred recently, including the addition of nitric oxide from supersonic planes and chlorine compounds from spray deodorants. Scientists have tried to estimate the ultimate effect of these additions on the atmosphere as well as the time it may take for changes to occur. These calculations are difficult because of the many complex catalytic cycles in the atmosphere. Different calculations have not always agreed, and we cannot tell yet from observations whether we are calculating correctly. This is some indication of how hard it is to estimate the time scale for the processes of chemical evolution. We believe this cannot be done with any degree of reliability at present and so must leave this important question open.

For life forms not based on reactions between molecules, the time scales involve quite different factors. Suppose that only the internal motions of individual atoms, rather than the chemical form of substances, change in a particular life process. It would be possible for this life to develop through an exchange of radiation among the atoms, rather than through collisions due to the movement of the atoms. This would tend to make evolution much more rapid, since radiation travels much more quickly than atoms under most conditions. One atom can influence another atom by the exchange of radiant energy much more quickly than by moving near enough to combine chemically with it. This could be especially important for life at low temperatures and low densities, where the rate of collision between atoms is so slow that chemical life does not have time to evolve.

Finally, we can say a few things about the time required for life processes in very high-density situations, such as those found in the deep interior of planets and stars. Here the main bottleneck is a lack of mobility of the atoms, or atomic components, which could become ordered. If the matter becomes solidified at high density, the atoms become almost immobile, and each atom can only react with a small number of neighbors. This situation might allow for a kind of life process which consists of a gradual growth of order outward from a central nucleus. Again, this would probably involve the internal constitution of each atom, rather than chemical combinations of different atoms. An example of such order also exists among lasers, which can occur as solid materials as well as in gases. Here again, the possibility of one atom influencing another by transferring radiation is essential.

High density does not always imply solidification. If the temperature is also high enough, where the concept of "high enough" varies a great deal from substance to substance, matter can remain fluid even when it is very compressed. For example, hydrogen gas is believed to remain fluid even at densities that are greater than those of ordinary solids. These conditions may exist in the deep interior of the planet Jupiter. Under such conditions the mobility of the atoms remains quite high, and there is no restriction on life due to slow reaction rates.

If there is a stable fluid form of matter at high density, order can develop in it. The rate at which order evolves is likely to be much faster than for familiar matter because the collision rate in high-density material is correspondingly greater. In some cases the densities can be millions of times greater than those of Earth matter, and in a liquid or gas of such density as may exist in a white dwarf star, the collision rates will be hundreds of times faster. Even higher densities and collision rates occur in neutron stars. High temperatures also increase collision rates because the atoms move faster. A likely result is that if life can arise in such conditions, it would do so rather quickly compared with the time it took to evolve on Earth. This is providential because the period over which a neutron star surface has a source of free energy available is probably measured in thousands rather than billions of years, and by the standards of Earth-life, neutron star life must be a rather fleeting phenomenon, somewhat as the life span of a mayfly appears to a human, or that of a squirrel would appear to an intelligent Sequoia tree.

The important question of how the time required for the development of life in some environments compares with the time over which that environment is stable is hard to answer. Fortunately, there are many environments where conditions are stable over periods of time comparable to the present age of the Universe, so that this question is equivalent to asking whether life in them has had enough time to evolve at all. But other environments, such as the atmospheres of some hot stars, have existed for a much shorter time than the Universe, and it is uncertain whether the time available has been enough for life to evolve in them.

Of all the conditions for life that we have discovered, the availability of sufficient time is the hardest to evaluate. Perhaps the most we can do is to indicate, for several types of environments that are widespread in the Universe, how the factors that we know are able to influence the rate at which life develops compare with similar factors on Earth. When these factors act to speed up the evolution of life, we can look to its early development in the particular environment. When they act to slow it down, we can expect life to develop slowly, if indeed it has had time to evolve at all.

Our survey of the prerequisite conditions for life leads us to the conclusion that a wide range of environments in the Universe may satisfy these requirements. While we have found that any specific kind of life may be restricted in its scope to special environments, this can be compensated for by the many varieties that can exist. Consequently, there are few environments which can definitely be ruled out as possible places for development of life. But this possibility is not the same as predicting that life exists in all these environments. In order to make such predictions, we need to examine specific environments in more detail and try to construct in our imaginations the kind of life that could thrive there. This will be the work of the following chapters. We will discuss forms of life based on molecular arrangements, which we call chemical life, and forms based on the ordering of other properties of matter, which we call physical life. Also considered will be specific habitats in our own solar system and elsewhere in the Universe, and we will try to match a number of them to the types of life that may occur there. We will not attempt to describe all possible types of **extra**terrestrial life, but if our speculations score a few hits among the forms of life that inhabit the Universe, we will be satisfied.

LIFE BEYOND EARTH

The universe is not only queerer than we imagine—it is queerer than we can imagine.

—J. B. S. Haldane

(We'll see about that.)

"Ammonia! Ammonia!"

Chapter 8

Chemical Life

The Fittest of All Possible Worlds?

> "I pray for one last landing
> On the globe that gave me birth
> Let me rest my eyes on the fleecy skies
> And the cool, green hills of Earth."

This poem from an early science fiction story by Robert A. Heinlein anticipated the admiration of our home planet many of us felt when we saw the first photos of Earth taken from space. The views of swirling clouds covering blue seas stood in striking contrast to the alien, barren rocks of Moon or the red glare of the Martian desert. The love we feel for our planet is natural and analogous to the appreciation we have for our own bodies. As we have extended our knowledge of the intricate workings of our cells and of the chemicals of which they are composed, we have tended to include those biochemical mechanisms in our feelings of pride.

We have recognized that these mechanisms are the product of a long evolutionary process, according to the Darwinian process, "survival of the fittest." They are particularly fit for their purpose, sustaining life.

Unfortunately, some writers have been tempted to extend the notion of fitness, along with our feeling of love and pride, to the Earth.

The notion of fitness of an environment for life was suggested by the American chemist Lawrence Henderson in

1912 when he published *The Fitness of the Environment.* This book, reprinted in 1958 with an introduction by George Wald, might be regarded as the Bible for this line of thought. It is essentially a hymn of praise to water and carbon dioxide (which, when dissolved in water, becomes carbonic acid). Some of its flavor can be obtained from the following excerpt:

> *The fitness of the environment results from characteristics which constitute a series of maxima-unique or nearly unique properties of water, carbonic acid, the compounds of carbon, hydrogen, and oxygen, and the oceans—so numerous, so varied, so nearly complete among all things in the problem that together they form certainly the greatest possible fitness.*

Henderson concludes that the particular fitness of these properties for life is a coincidence too remarkable to be ignored. Perhaps the properties have been preselected in some way and represent a mysterious cosmic purpose. Henderson's disciple, George Wald, put it quite succinctly, "Sometimes it is as though nature were trying to tell us something, almost to shake us into listening."

This overall view, including purpose, is of course related to the predestinist view discussed in an earlier chapter. Its particular application to the fitness of the chemicals and conditions of our biosphere for life on Earth warrants an extra analogy: A geographer with a predestinist viewpoint might eventually be struck by just how fit the Mississippi River was for its valley. It flows in exactly the right direction, with exactly the needed contours and tributaries, to ensure the drainage of the waters of the central United States into the Gulf of Mexico. In doing so, it passes conveniently by every wharf and under every bridge in its path. The geographer might then attempt to replace the Mississippi, hypothetically, with the Amazon River. Superimposing the Amazon onto a map of the United States, he might notice at once that it flows west to east. This would not work, as it would have to flow over mountains. Even when he turned the river in the "right" direction, he would notice many difficulties. New Orleans would be flooded by the large Amazon delta, and an endless number of roads and towns would be submerged. He would conclude that the Amazon was unfit, and the Mississippi eminently fit, for its purpose.

Let us restrict the situation further. Assume that the geographer had no knowledge of other river systems but had studied the Mississippi extensively. He would notice that any major change in the river's shape would cause damage and dislocations, and conclude that this shape was the only one possible for a functioning geological system. If other rivers existed, they must have the same general shape. In an analogous vein, George Wald has concluded, "For these and similar reasons I have become convinced that life everywhere must be based primarily upon carbon, hydrogen, nitrogen, and oxygen, upon an organic chemistry therefore much as on the earth, and that it can arise only in an environment rich in water."

We have already given a name to this point of view, which would dismiss out of hand other possible bases for life—carbaquist. It is even more absolute in its assumptions than the predestinist viewpoint. The predestinist believes that life on Earth is the way it is because Nature has arranged the game that way. The dice are loaded. Nevertheless, it might be possible, through human intervention, to create other novel forms of life that would not have evolved "naturally." The carbaquist rolls dice that are not merely biased but rather have all sixes on one cube and all ones on the other. No outcome, other than a seven, can even be conceived.

In the absence of any example of a life system other than the one we know on Earth, the carbaquist view cannot be refuted conclusively. This situation puts us in the position of an intelligent, isolated, and musically naïve tribe who has been given a violin by a passing traveler. He has demonstrated its use but told us nothing about the principles of music. We might toy with it and attempt to replace the strings, bow, sound post, or pegs with other materials. When these efforts gave bad results, the carbaquists in our tribe would then declare firmly that no other basis for the production of music could exist. Those who discussed the possibility of cellos or even guitars would be looked upon with disdain. This state of affairs would continue until another traveler presented the tribe with a trumpet or an organ.

We do not have the trumpet or organ at this time. What we can do is consider the arguments of the carbaquists in greater detail. When we see that the presumed barriers to the existence of other bases for life do not exist, we will be prepared to accept the possibility of other life forms.

WATER

Water, the essential fluid of Earthlife, is composed of two of the three most common types of atom in the Universe: hydrogen and oxygen (the third, helium, has no known compounds at all). Water will occur in many different environments, though not necessarily everywhere. Moon, for instance, appears to have no significant amount. Water has been shown to be a suitable medium for life on at least one biosphere, our own. It can exist in liquid form over a broad temperature range at suitable atmospheric pressures. In those locations where it is a plentiful liquid, it is reasonable to expect that it will be used by existing life forms, if the other requirements, such as a flow of energy, are provided. These facts do not establish that another solvent may not serve as a basis for life when liquid water is absent. The carbaquists, however, argue that the special properties of water make it unique as a support for living organisms.

One of the special properties of water is its great ability to serve as a reservoir for heat. A given quantity of water requires more heat to melt, warm up in temperature, and boil than does the same amount of most other chemicals. A given amount of heat produces a lesser change in the temperature of water than most other liquids. Bodies of water thus tend to stabilize the climate of their environment. On Earth a location by the sea will often have smaller temperature swings from day to night than one a few miles inland. This property may be pleasant, but it is hardly essential to life. Other climatic features on the planet being investigated could serve to keep temperatures steady, as happens on Earth. San Francisco and New York are both on the sea and at about the same latitude, but the climate of San Francisco is much steadier throughout the year. At other locations on Earth, the climate is sufficiently warm throughout the year that water never freezes.

A more important consideration is that swings of temperature, even those leading to the freezing or evaporation of the liquid medium, need not be destructive to life if they are not permanent. Organisms on Earth have adapted to survive being frozen in ice, or exposed in deserts in which the only liquid water present is the protected supply within the interior of the organism. The most important consideration for survival is that the vital chemical structures of the organism be stable at the highest temperatures to which it is exposed.

Another property of water that is greatly admired by the carbaquists is its polar character. This term means that each molecule of water has a distribution of electrical charge associated with it with definite positive and negative ends. Water is then a very good solvent for other polar substances such as salts, and an appropriate medium for the conduct of reactions involving charged molecules such as acids and alkalis. Because of the polar nature of water, many substances that dissolve in it become separated into positive and negative ions. We have seen that this may speed up or slow down the rate of chemical reactions between the dissolved substances. Also, the molecules in solution are free to move about and react, which they would not be free to do if they were in solid form. Life on Earth makes good use of this polar nature of water, and many crucial biological molecules, such as proteins and nucleic acids, carry electrical charges. But other factors such as temperature also influence the rate of chemical reactions. There is no reason to believe that life processes, such as storage of information, catalysis, and replication, could not be carried out with uncharged molecules in another solvent. We will consider this further when we discuss hydrocarbon life.

Lawrence Henderson considered water to be an unreactive substance: "Water is really, at the temperature of the earth, and in comparison with most other chemical substances, an extremely inert body." The experience gained during an additional two-thirds of a century has changed our view on this matter. As stated in an article in *Chemistry in Britain* by Felix Franks, the editor of *Water, A Comprehensive Treatise,* "It is in fact one of the most corrosive substances known." The apparent inertness of water stems from the fact that it has been on Earth for a long time, and has completed its chemical reaction with the geological substances susceptible to it. The violent effect of small amounts of water on the surface of Mars, which will be described in Chapter 9, puts Henderson's statement in its proper perspective.

Among the substances that are corroded by water are our key biological macromolecules. They are assembled in a way that involves the removal of water from their component parts. The water in our cells is continually trying to reverse the process and dismantle the large molecules into their components. The damage that is produced is particularly troublesome in the case of DNA, and living cells on Earth maintain themselves by continuously repairing the wounds inflicted on DNA by water. Not all the properties of water are fit for life.

The attribute of water that is perhaps the most highly valued by the carbaquists is that of expansion on freezing. This *is* a fairly unique property. One consequence of this ability is that a pond of water starts to freeze at the top, and the layer of ice which forms protects the remainder of the pond from further solidification. Much has been written about the results to Earth if water did not have this property, in a kind of reasoning akin to that involved in replacing the Mississippi River with the Amazon. Water does, however, have this characteristic, and therefore the Earth is as it is. For the purposes of this book it is more important to ask whether life could exist on a planet where another liquid existed and where its ponds and lakes tended to freeze from the bottom up. We have already mentioned several facts which indicate possible solutions. The climate might be such that the ponds never froze, or the organisms in frozen ponds might hibernate until spring returned.

To conclude, let us quote Dr. Franks again: "Water was on this planet long before the evolution of life, and therefore the processes which make up life . . . have had to make use of the rather eccentric properties of this natural substrate." It is not that water is essential to life. Rather it is that the life we know has evolved in circumstances where it had to adapt to the properties of water. On another planet the particular properties of other solvent systems could be used by the life that evolves there, and this life would adapt to these other solvents.

A number of writers on extraterrestrial life have prepared lists of solvents that could serve as substitutes for water. If anything, the lists have been too conservative. They have featured substances such as ammonia and hydrocyanic acid, which are polar and can serve as effective heat reservoirs. As we have pointed out, these properties are not essential. Some similar solvents, such as liquid hydrogen fluoride, have been set aside on the grounds of low cosmic abundance. It may be unwise to do this. Fractionation processes exist in nature that may greatly enrich the amount of an unusual element in particular locations. In the Universe as a whole, for example, 87 percent of the atoms are hydrogen and only 0.002 percent are iron. Iron is, however, the most abundant element in the whole planet Earth (35 percent), while hydrogen is present at less than 1 percent. Gold and mercury are in low abundance on Earth as a whole, but are concentrated many thousandfold at special sites.

Some solvents have been disregarded because the range of temperature in which they are liquid (at Earth pressures) is narrow. However, boiling points can be raised by higher atmospheric pressures, and freezing points lowered by the presence of dissolved substances. Furthermore, the crucial factor that determines whether a substance on a given planet remains liquid over a whole year is not the difference between the boiling point and the freezing point. Instead, it is the ratio of the absolute temperatures at which the liquid boils and freezes, and that ratio is higher for several other liquids than for water. The most important properties that a solvent must have for life are that it dissolve enough substances to form a self-organizing system, and that its elemental composition be relatively simple (to increase the probability of its formation). Substances that can exist as liquids, under conditions where water is not liquid, should particularly command our attention. For example, ammonia, hydrocyanic acid, methane, and hydrogen fluoride have low freezing points. Sulfur (which may exist in liquid form beneath the crust of Jupiter's moon, Io) and concentrated sulfuric acid (present as droplets in the clouds of Venus) have relatively high boiling points. Many metallic elements are liquid at even higher temperatures. There is no shortage of candidates for solvents in which life could arise in other environments.

CARBON

The second great set of pillars that support the temple of carbaquist beliefs are the special and indispensable properties of carbon. To quote Norman Horowitz, Professor of Biology at the California Institute of Technology: "The capacity of generating, storing, replicating, and utilizing large amounts of information implies an underlying molecular complexity that is known only among compounds of carbon."

Let us give carbon its due. Carbon atoms have the ability to form bonds to four other atoms at one time (i.e., they are tetravalent). This property permits the construction of an enormous number of molecules containing carbon that have all types of sizes and shapes. Other elements can also form four bonds, however; for example, silicon, a principal ingredient of the rock on this planet. Some elements that often form three bonds—nitrogen, phosphorus, and boron —can on occasion have four. Sulfur normally forms two bonds, but is known to bond to as many as six atoms in certain compounds. It is not even clear that it is necessary to have tetravalent atoms to form complex structures. Atoms that form one bond can only link up in pairs. Those that form two bonds can create chains. But once a third bond is made to each atom, then complex networks become possible. Carbon is certainly not unique in its ability to generate molecular complexity.

Another property of carbon cited by the carbaquists is its ability to form stable bonds to itself. It is therefore possible to make large molecules out of only two elements; for example, carbon and hydrogen. (If only carbon atoms are used, one gets an endless connection of bonds, as in a diamond.) Other elements can form stable bonds to themselves, but at our familiar temperatures on Earth they do not do it as well as carbon. However, it is not clear why it is important that the backbones of the large molecules used in chemical life be made of just one element. It would be perfectly adequate for them to be constructed of two or more alternating types of atom, so that no atom would bond another similar atom. In fact, partial alternation occurs in the chains of atoms that make up the most important large molecules of Earthlife. The proteins have only two carbons, then a nitrogen, then another two carbons, then a nitrogen, and so forth, as the repeating unit of their backbones. The nucleic acids have three carbons, then oxygen, phosphorus, and oxygen, as a unit that is repeated many times. (See Fig. 20.) The ability of carbon to form long chains by itself is not a property that Earthlife takes advantage of in its most important molecules.

a. $-C-C-N-C-C-N-C-C-N-$

b. $-O-P-O-C-C-C-O-P-O-C-C-C-$

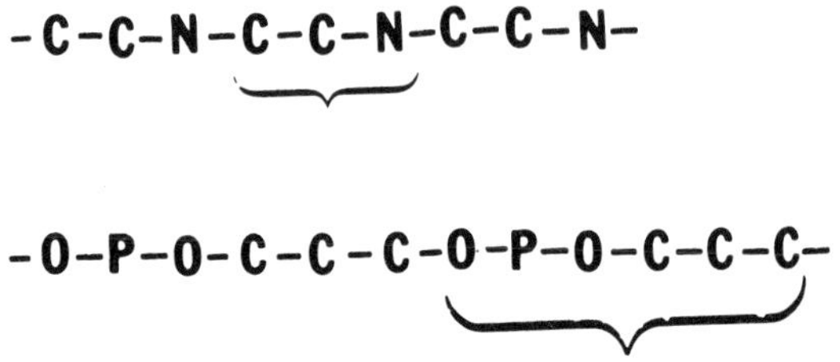

Figure 20. The arrangement of atoms in the chains that make up the backbones of (a) proteins and (b) nucleic acids. The brackets indicate the repeating units.

Carbon is not unique as a basis for generating molecular complexity, but it is very effective. The number of known organic compounds has run into the millions. We have seen that the complexity of carbon chemistry may have been a severe handicap, rather than an advantage, in the origin of life on Earth. Only a limited number of small organic molecules, certainly less than a thousand, play an important role in the general metabolism of Earthlife. The larger molecules are made by joining together a selection of the smaller ones. It may have been quite difficult for these vital molecules to locate each other and form an organized state in the presence of enormous quantities of irrelevant ones. Amusingly, the predestinist view might be more appropriate for the origin of a type of life based on elements that can form only a small number of distinct compounds.

It is quite conceivable that a basis for life could be constructed using an alternative chemistry in which the possibilities are not as vast. We can compare the situation with the problem of storing information in printed form. To accomplish this, the English language uses twenty-six capital letters, twenty-six small letters, a number of punctuation marks, and spaces. Using about sixty characters in all, an enormous amount of information can be stored in a line of text. The same amount of data can be stored by a computer, using only the two characters 1 and 0. The price that must be paid for this simplification is that six lines are needed to hold the contents of one English line. Just the same, the information is stored. In the same way, a less complex chemistry might serve as a basis for life, perhaps with a larger number of components needed in each cell or unit.

One carbon-containing molecule that receives particular praise from the carbaquists is carbon dioxide. It exists both as a gas in the air and dissolved in water, and thus facilitates the exchange of carbon between different parts of the biosphere (see Chapter 5). This is again a pleasant property but not an essential one. Phosphate, which is vital to Earth-life, circulates indefinitely in the biosphere without entering the atmosphere.

Carbon is relatively abundant in the Universe, and its compounds are stable and suitable for life up to a temperature of hundreds of degrees centigrade. There have been occasional arguments as to whether life could be constructed using silicon or another element in the place of carbon. It is unlikely, however, that if carbon is present somewhere it would be excluded in the evolution of a life form just to settle academic arguments on Earth. We would expect to encounter carbon in many or most extraterrestrial life systems based on molecular arrangements, but its role may be less dominant than it is on Earth.

With the carbaquist assumptions put into proper perspective, we can now consider some possible alternative chemical systems of life. But perhaps we should pause first to consider the opinion of Professor Horowitz on previous proposals of this type: "It is not possible to evaluate these proposals because they have not been developed in detail. In the absence of detailed models, all such speculations must be viewed with skepticism."

This view is to some extent an outgrowth of an attitude held by some biologists, a distrust of theoretical speculation as opposed to a methodical analysis of experiment and observation. Such distrust is not the rule in other branches of science. In physics and chemistry some of the greatest advances have come from speculative leaps beyond what could be directly justified by experiment. This often happens when predictions of new phenomena are involved, which is much more common in physical science than in biology. Especially in a situation where we possess only a small part of the relevant facts, a reliance on theory to guide us is essential. To paraphrase Einstein, theory without experiment is lame, experiment without theory is blind.

We can imagine an opinion similar to that of Horowitz being expressed in the Spanish court in 1491, concerning the existence of undiscovered continents or sea routes: "No such speculations can be considered seriously until detailed maps of the routes and lands to be discovered are presented." In fact, the discovery of new life forms would be quite analogous to the discovery of the Americas. In both cases no hard proof of the discoveries existed in advance. However, the existence of alternate life forms would violate no law of biology, as the discovery of new continents violated no law of geography or geology. In both cases there was more than ample space in which the areas to be discovered could exist. The situation is very different from attempting to prepare a perpetual motion machine, or accelerating an object from rest to speeds exceeding those of light. Such projects would try to overturn existing laws of science. The discovery of alternate biologies would violate none.

Christopher Columbus did not have a detailed map when he set out on his first voyage of discovery. He did have a direction, however, and permission to sail.

The arguments that we have given thus far have been designed mainly to clear away the misleading views of the carbaquists and predestinists about the materials essential for life. In the remainder of this chapter we will take a more constructive approach and suggest some specific alternatives. First, we will consider more conservative possibilities, in which the chemistry remains that of carbon and water but other materials replace protein and nucleic acid as the fundamental macromolecules of life. These possibilities could be relevant to other Earthlike planets, in which chemical evolution took a somewhat different course in its very early stages. Then, we will consider more exotic possibilities, in which water or carbon are not as fundamental as in Earthlife. These might be relevant to an environment for which the temperature range is still such that chemical reactions are prevalent, but where different materials are the preferred ones for life.

ALTERNATIVES TO
PROTEINS AND NUCLEIC ACIDS

An extreme form of carbaquism holds that only nucleic acids and proteins, among organic molecules, can carry out the genetic and catalytic functions necessary for life. Even the more complex structures found in cells, such as ribosomes and membranes, may be unique solutions to tricky technical problems. Thus, no other type of life save the familiar nucleic acid-protein combination could exist here or elsewhere. Such a view might also be held by the predestinists, who believe that this type of life is an inevitable consequence of natural processes on Earthlike planets.

If we accept a different viewpoint, that of gradual chemical evolution, then very different possibilities open up for life on an Earthlike planet. We know, of course, the outcome of evolution on Earth, and there is no way that we can set the clock back and rerun the experiment. It is possible, though, that some centuries from now humans may have the chance to explore a similar planet in another solar system. If life exists there, it would be reasonable for it to be based on water, which is the one liquid that is abundant on a planet of the same size, temperature, and atomic composition as Earth. Carbon is a versatile element, and we might expect it to be utilized if it were as readily available as it is on Earth. A number of the simple organic substances that occur in our biochemistry might also find place in another system. Proteins and nucleic acids, however, are large, specialized molecules. An alternative turn of events on another planet might well have changed the course of evolution so that other substances serve for the functions of catalysts and information storage.

Of the above two types of large molecules, proteins might be more likely to occur. Their component amino acids are formed under a number of circumstances (though not inevitably). If amino acids were somehow concentrated by natural processes, their combination into proteins would be plausible. However, other solutions to the catalysis problem exist. Many chemical compounds have an intrinsic ability to catalyze selected reactions. Chemists have had reasonable success in designing small molecules that mimic the effects of enzymes. The greater catalytic power of enzymes resides in the greater complexity of their structure.

Could alternative large molecules of the size of enzymes be constructed which rival them in their catalytic power? Undoubtedly yes, though this has not been done as yet. Organic chemists appreciate that a vast array of small molecules can be connected in various ways to form larger ones made of repeated links. On the prebiotic Earth, the combination of such small molecules into larger ones was as likely an event as the union of amino acids to form proteins. Many substances in our everyday lives, such as nylon and cellophane, are macromolecules that have been made by joining smaller units. These substances are highly regular in their molecular structure, however. They lack the specified irregular arrangement of markers that are characteristic of proteins and are vital to their catalytic power. (See Fig. 21.)

In the laboratory it would be possible to construct large polymers which are irregular arrangements of simple units other than amino acids. It is also possible that such irregular

Figure 21. A symbolic representation of the construction of nylon. It is made of two alternating units, indicated in dark and light color.

polymers were formed naturally during prebiotic evolution, both on Earth and in other environments. Most such "random" polymers would have no special catalytic properties and so would not enter into chemical cycles. The labor involved in producing them in the laboratory at present would be enormous. However, chemical theory is not yet in the state where it can predict accurately in advance exactly which giant molecules would have interesting catalytic properties. There would be a considerable risk that the labor invested in the first efforts would yield no positive result. Understandably, there has been no rush of chemists to enter this area of research. The possibility that an alternative set of catalysts could be produced by natural selection on another world remains a very real one, however.

Much less is known about molecular information storage and the principles involved in the copying of molecules than about catalysis. The nucleic acids were discovered in 1869 by Friedrich Miescher, but it was not until 1944 that the first serious suggestion was made that these substances were the genetic material. Only in 1953, when detailed information about the actual structure of DNA was available, were the actual features of its genetic function grasped by James Watson and Francis Crick. Disputes raged over the details of their proposal for years. Even today a small group of dissidents maintains that the two chains of DNA lie side by side in nature rather than intertwined in a double helix, as stated by Watson and Crick.

A crucial feature in the Watson-Crick theory of the replication of DNA is the scheme by which a particular subunit on one chain of DNA "recognizes" (forms chemical bonds in a highly selective manner) a different subunit on the other chain of the double helix. The bonds which are used in this recognition scheme are hydrogen bonds that are weaker than the usual connections that hold organic molecules together.

Isolated subunits of DNA do not form hydrogen bonds to each other when they are dissolved in the same water solution. They prefer instead to bond to the water molecules. It is only when many subunits are brought together to form two DNA chains which match each other in terms of the Watson-Crick recognition scheme, that the power of all the hydrogen bonds acting together becomes sufficient to keep the two chains united. It is somewhat like the case of a thick rope which gets its strength from the combination of many weak, thin threads.

The scheme in which the two components of the double helix come apart at an appropriate moment and are copied, eventually to produce two identical double helices, is the basis of replication. It is important to remember that the nucleic acids cannot carry out this procedure alone, but need the assistance of a number of enzymes.

A critical point that emerges is that nucleic acids do not fill their genetic role by the use of any unique property that they alone can possess, but operate through the application of a number of conventional chemical interactions in a very involved process. The specific details of their function have been understood only after a long study of the process itself. It has not been done by theoretical deduction. It is quite likely that many schemes can exist by which other molecules can catalyze their own replication. The functioning of these alternative replicating systems will probably be as complex as that of our own nucleic acid-protein system. It is not likely that we will learn of their existence by theoretical deduction; they must be discovered either by the exploration of the Universe or by laboratory experimentation.

The labor involved in laboratory efforts to prepare alternative replicating molecular systems, like that needed to prepare other types of macromolecular catalysts, would be enormous. A number of ingenious proposals have been made, however, by the Glasgow chemists, A. G. Cairns-Smith and C. J. Davis. They described a number of organic molecules which, while differing considerably from DNA in structure (some examples lack both nitrogen and phosphorus), could still fulfill a genetic function. In some of the proposed examples, molecular recognition operates through hydrogen bonds, as it does with DNA. In other cases, however, stronger chemical bonds or electrical attractive forces are used. These possibilities may or may not work if we try to construct them. The likelihood that alternative genetic systems to DNA can exist is a strong one, nonetheless, and they may actually be functioning at this time on other worlds.

There are no very Earthlike worlds in our own solar system, and it is likely to be centuries before we can search for such planets in orbit about other stars. The environment on other worlds in our solar system is very different from that on Earth. If life with a different chemical basis exists on them, or there is no life at all, this result will be attributed to the difference in environment. The question would remain open whether nucleic acid-protein life is unique for Earth-type planets.

EXOTIC LIFE ON EARTH?

There is, however, one planet close at hand, which has been insufficiently investigated and which on closer inspection may provide us with information about alternative forms of life. It is the Earth itself.

This statement may appear to contradict what we said earlier in the book. We will reaffirm that all *known* life forms on this planet operate within the nucleic acid system. It is not yet clear that life forms of this type represent the entire living contents of the planet.

As we stated earlier, the extensive biochemical similarities between all known organisms on Earth suggest that they have evolved from a common ancestor. This ancestor was itself highly evolved. No trace has yet been found of divergent life forms which may have existed before the common ancestor. It seems likely that the development of some feature, such as the storage of information in DNA or the perfection of the genetic code, gave one type of life such an advantage that it has come to dominate the biosphere. One indication of this dominance is that nucleic acid-protein life has proved to be quite versatile in occupying a variety of habitats on Earth. In the test tube the macromolecules themselves are quite sensitive to the effects of heat, acid, or alkali. They readily lose their activity under these conditions. Yet organisms have developed that can defy these limits. Certain strains of microorganisms have proved to be particularly adaptable in this respect. Bacteria and algae grow in hot springs of near-boiling water, in strongly acid or alkaline natural pools, and in artificial media that should destroy their key molecules. The microorganisms protect themselves from acids and alkali by pumping these substances out of the cells. They do not do the same thing with heat, but rather make minor modifications on their nucleic acids and proteins which give added heat stability.

The cells of most organisms on Earth are quite vulnerable to dehydration. This can occur directly in dry environments, or when the cells are immersed in concentrated salt solutions. In the latter process, water is drawn out of the cell into the solution, leading to dehydration inside. A variety of bacteria, called halobacteria, have learned, however, to thrive in concentrated salt solutions. A necessary modification has been made on their cell membranes. Some strains of halobacteria may then die if put into more dilute salt solutions. The effects of dehydration due to low humidity can be countered by some organisms by the development of impermeable cell coatings. A more dramatic response is illustrated by the tardigrade described in Chapter 6. Under dehydrating conditions, this tiny animal goes into a dormant, contracted, inert state in which up to 97 percent of its water is lost. When water is restored after months or years, the animal returns to activity.

The adaptability and the limitations of nucleic acid-protein life can be illustrated by considering the cold windswept deserts of Antarctica. Upon visual inspection, this environment appears totally barren, and it is considered to be among the most hostile to life of all locales on Earth. However, bacteria and algae were recently discovered residing within the rocks. They exist inside light-colored rocks, on the side that faces the Sun, and at a depth that protects them from the cold but allows them access to sunlight. They metabolize actively when temperature conditions are favorable and remain inactive at other times.

It can be deduced that nucleic acid-protein life has extended itself fully in utilizing marginal locations in the biosphere, such as the interior of the Antarctic rocks, but that there are still limits that cannot be exceeded by minor adjustments in the system; for example, the rock exteriors and the surrounding soil. Such limits may be established by climatic extremes or, more subtly, by deficiencies in key nutrient elements, such as sulfur or phosphorus, or iron and magnesium.

Environmental niches of these types constitute possible refuges for alternative life forms. A form that flourished before the takeover by nucleic acid-protein life could have possessed some advantage which allowed it to survive in these marginal locations. A plausible analogue is provided by the methanogens. It has been suggested that they flourished before the advent of an oxidizing atmosphere for the Earth. Unable to adapt, they maintain their existence in anaerobic locations such as the floor of the Black Sea. Recently, it has been demonstrated that the methanogens are related to both the halobacteria and the bacteria that inhabit warm acidic springs. It has been proposed that they be placed in a separate kingdom, the archaeobacteria. While still definitely members of the nucleic acid-protein system, there are sufficient differences in detail in the archaeobacteria to convince some scientists that they represent an ancient divergence from the main lines of evolution.

This version of the history of the methanogens suggests another generality about the biosphere: that its activities have gradually reconstructed Earth's environment to make it a better place for some kinds of life and a worse place for other kinds. Individual living things no longer need to hide deep in the seas to protect themselves from ultraviolet light, as the thin stratospheric ozone layer does this for them. However, a life form that needed to use this ultraviolet light for its metabolism would not appreciate the change.

A much more exciting discovery than methanogens would be that of survivors of an even more ancient divergence, perhaps one that preceded the establishment of a nucleic-acid hereditary system! Such a discovery would represent a far greater biological novelty than the Loch Ness monster, and provide a much more significant missing link in evolution than those that have been sought in studies on the origin of the human race.

Is it possible that a novelty of this magnitude has been overlooked to date by microbiological science? Yes. To cite a recent textbook on microbial ecology,

> *Yet the species composition of many environments has been inadequately described, and evidence for entirely new and unusual microorganisms is still being obtained. . . . Scores of extreme environments are known on the surface of the globe, but just a few have been examined, and only in rare instances have the studies been extensive.*

Microbiologists cannot at present determine the chemical nature of a new species of microorganism, using only a single specimen which can be observed under the microscope. It is necessary to allow the strain to multiply by growing it in a suitable culture medium in order to obtain enough material for analysis. In studies on the types of microorganisms present in even such conventional sources as pond water and soil, the growth of certain species may be inhibited by the presence of competitors that develop faster in the laboratory, or by unusual nutritional needs such as a concentration of organic compounds lower than the one used in conventional culture media. When a highly unusual organism is involved, it is even more likely that it would fail to develop in media designed to enhance the growth of conventional forms of Earthlife. As we will see, this problem has been crucial in designing experiments to detect life on Mars.

An additional impediment to the discovery of alternative life forms on Earth may be the reluctance of many microbiologists to consider seriously the possibility that such forms exist. Conservative scientists sometimes state that speculative topics of this type should be reserved for cocktail party conversation. In our experience, some microbiologists have tended to tiptoe away even when the topic was raised over cocktails! One investigator who did not back off, however, was Joshua Lederberg, who won the Nobel Prize for his work in bacterial genetics. In conversation, Professor Lederberg agreed that non-nucleic acid life might exist on Earth, not only in remote niches but perhaps right under our noses. He had once considered cultivating microorganisms in the presence of sufficient radioactive phosphate to kill microorganisms using nucleic acids as their hereditary material. Alternatively, one might try to cultivate organisms in media rigorously deprived of phosphate. Would anything else grow then? It would be exciting to find out.

A search for an alternative life form on Earth would run a considerable chance of failure even if the target should exist. There is no clear guideline where or how to search. The methods tried, such as the ones listed above, might still fail to allow the growth of the desired organisms, through a lack of knowledge of the proper nourishing medium. However, the scientific reward is sufficiently large to make such an enterprise worthwhile. Quite recently, a billion dollars was spent on the Viking project on Mars, with the search for new life forms as a principal goal. The same objective could also be pursued on Earth at a tiny fraction of the cost. Some of the manipulations involved are simple enough that amateurs, and ambitious high-school students, could participate actively. An ingenious idea, and some luck, could lead to a discovery of momentous significance in the history of biology. Is anyone out there interested?

ALTERNATIVE CHEMISTRIES

If alternative chemistries based on carbon and water can serve as a basis for life on Earthlike planets, then the possibilities for life in the Universe will be greatly increased. A still greater increase will result, however, if life can exist on planets very different from Earth, as the number of such worlds is likely to be far greater than the Earthlike ones.

Could chemical life exist on worlds where water is unavailable, or where organic compounds break due to high temperatures? There are many possible substitutes for water and carbon in such environments. We present three of them, in the form of two hypothetical worlds where water is replaced by another medium in which living processes take place, and one world in which living things are made of materials other than carbon compounds.

1. Life in Ammonia

It is a spring day on Lake Ammon, the largest lake on the imaginary planet Frigidus. A brisk wind churns up waves on the lake. The Sun, smaller in its apparent size than the one we are accustomed to on Earth, shines dimly through wispy clouds. Its light reflects off the hills of ice that encircle the lake in the distance. Organisms of various types swim actively in the lake, devouring each other as well as the plant forms that use the sunlight as a source of energy. The creatures are not at all inconvenienced by the temperature, which has risen to a brisk −53° C. This temperature, which on Earth we find only in Siberian or Antarctic winters, is mild and benign to these organisms. They are also quite happy with the composition of the lake, which is ammonia containing some dissolved water. The tissues of the creatures that inhabit Lake Ammon are bathed internally in the same fluid as the lake. It is liquid over the whole temperature range that occurs on Frigidus. The planet has a dense atmosphere of nitrogen. It rotates rapidly on its axis and revolves in a very circular orbit about its star, which accounts for the relatively constant temperature on its surface. The conditions on Frigidus are idyllic to the creatures in the lake, though they would be lethal to organisms from Earth.

This pastoral scene would not please the carbaquists, who would deny it as a biological possibility anywhere in the Universe. Ammonia would not be considered to be suitable as a solvent for life, despite the fact that many of its properties resemble those of water. It has a good capacity as a reservoir for heat. It is reasonably polar. It dissolves salts and supports reactions between charged substances, such as acids and alkalis. Professor Wald, however, is not satisfied: "Ammonia ice sinks in liquid ammonia repeatedly and unequivocally, hitting the bottom of the vessel with a distinct thud." As we have already pointed out, however, Lake Ammon does not freeze. Or if it did, the creatures in it need not perish, with the possible exception of a few unfortunates who may be crushed under falling cubes of ammonia ice.

A seemingly more potent objection to the existence of life in Lake Ammon may be made on grounds of temperature. As we discussed in Chapter 7, a specific chemical reaction will certainly proceed more slowly at −53° C than at +20° C. Professor Wald has written, "The processes that led to the origin of life within perhaps a billion years upon this planet might then take some 64 billion years in an environment of liquid ammonia." If the planet Frigidus were about as old as the Earth, these processes would not have progressed very far by now. Which is probably just as well, since the organisms they evolved would perish in Lake Ammon. However, the extent to which a given reaction changes with temperature varies quite sharply from reaction to reaction. The key quantity is called the activation energy. The larger the activation energy, the greater the change in the reaction rate for a given change in temperature. Many common chemical reactions on Earth have an activation energy in the vicinity of twenty [the scientific units used are kcal/mol (kilocalorie/mole)]. Such reactions would slow down not 64 or 100 times but 100,000 times in being cooled from 20° C to the temperature of our lake. On the other hand, a reaction with an activation energy of six would slow down by a factor of only thirty-two, for the same temperature change. In making his estimate, Wald has arbitrarily used an activation energy relevant to the chemistry of Earthlife, rather than a smaller one that could be relevant to chemical reactions in liquid ammonia. He has also used the wrong formula to compare the rates of chemical reactions at different temperatures.

We can get better guidance by noting that some Earth bacteria live and grow at temperatures of 93° C. Under those conditions one might expect that chemical reactions would proceed more rapidly, and that the bacteria would metabolize much more quickly than normal bacteria. While the bacterial generation times are shorter at higher temperatures, the differences are not striking and are less than would be predicted from a naïve application of the rules describing how chemical reaction rates vary with temperature. The "hot" bacteria may use enzymatic processes with higher activation energies than those in normal bacteria, and these reactions remain in step at the higher temperatures rather than at 20° C. In the same way, chemical reactions with low activation energies could proceed rapidly even at the temperatures at which ammonia is liquid. In short, the key to developing a suitable life chemistry at a given temperature lies in selecting chemical reactions suited to that temperature. As certain chemical processes, for example those involving unstable molecules called free radicals, have very low activation energies (often two to four), it is clear that chemical systems, and life, could function at temperatures even lower than those on our hypothetical planet.

New types of biochemistry would be needed in life conducted in liquid ammonia. Some types of compound used in Earthlife, such as fats, would not appear, as they would decompose in ammonia. On the other hand, the lower temperatures and abundance of nitrogen would open up new possibilities; for example, chains of nitrogen atoms. At any temperature in the very wide range over which molecules can exist, there are some that are stable and others that react vigorously. For a compound to be of use in life processes, it must be reactive enough to take part in metabolism, but stable enough to resist spontaneous decomposition.

Compounds with chains of carbon atoms, or mixed chains with carbon and nitrogen atoms are widely used in Earthlife. Proteins, for example, have in their backbones a repeating unit of two carbons, then a nitrogen. The carbon-to-nitrogen bonds are comparable in stability to carbon-to-carbon bonds. Bonds from nitrogen to nitrogen, however, are weaker and occur rarely in Earthlife. Chains made of nitrogen atoms are explosive at the temperatures we usually encounter on Earth. At lower temperatures, however, they would be more stable and suitable for constructing complex molecules (conversely, as the British scientist V. A. Firsoff has pointed out, one could use sugar to blow up bridges on the hot side of the planet Mercury). Nitrogen normally has three valences, but on occasion it will bind to four atoms as carbon does. It could therefore serve as a key building element, along with carbon, in the construction of large molecules that are as complex as proteins and nucleic acids but which function best in the low-temperature, ammonia-rich environment of Frigidus.

Life forms designed on this basis may actually exist within our solar system; for example, on a satellite of an outer planet. If no seas of ammonia exist in our solar system, however, then our encounter with ammonia life must wait until the day when we are able to travel to other stars.

2. *Life in Oil*

The seas of the planet Petrolia appear at first glance to be covered with a gigantic oil slick. In fact, they are composed *entirely* of oil slick, or, more accurately, a mixture of hydrocarbons of lower boiling point than those usually carried in oil tankers on Earth. The composition of these seas changes as the seasons wax and wane, as lighter boiling components evaporate or return in an oily rain. Petrolia may have started with no seas at all, but rather with an atmosphere rich in methane, which boils at −182° C at Earth pressures. Several of the outer planets or satellites in our own solar system may have methane-rich atmospheres. Chemical reactions driven by solar radiation would gradually have converted the methane to more complex molecules on Petrolia, some of which could remain liquid even at several hundred degrees centigrade.

Although oil slicks are quite toxic to a variety of Earth creatures, the seas of Petrolia have nurtured their own indigenous life forms, with a hydrocarbon fluid bathing their tissues and the contents of their cells. These organisms may utilize the solvent mixture prevailing in the oceans, or, if they choose, select a different mixture of the components for their own internal purposes, just as the salt content within marine organisms on Earth may differ from that of the sea or lake in which they dwell. It is also possible that the hydrocarbon beings, over the years, have modified the chemical composition of the seas in which they dwell, just as Earth organisms have converted our own atmosphere to one rich in oxygen.

Our scenario again will not please the carbaquists. If ammonia life makes them impatient, then the prospect of hydrocarbon life would cause them to grind their teeth to powder. We will freely admit that many salts, acids, alkalis, and compounds important to our own chemistry will not dissolve readily in this medium. Few processes involving highly charged intermediates will take place readily. Does this mean there would be such a dearth of chemical possibilities that no system of sufficient complexity for life could be organized? By no means. Such a conclusion would go against the experience of generations of organic chemists. These scientists, when faced with the task of assembling complex organic molecules without the aid of enzymes, have generally preferred to carry out their reactions in organic solvents rather than in water. Organic solvents are suitable for many sensitive, specific processes which would be ruined by the great reactivity of water. Ironically, one area in which water has been generally shunned as a solvent is the chemical synthesis of nucleic acids, the key molecules of Earthlife. The solvent of choice for this process has been pyridine, a substance made of carbon and hydrogen and having one nitrogen atom. If our chemical experience were the sole guide, the most likely place in the Universe to expect a prebiotic synthesis of nucleic acids would not be Earth with its vexing seas of reactive water, but a planet with oceans of pyridine. We do not wish to give undue encouragement to such a search, for the prospects of success are not very encouraging. Pyridine is a fairly complex molecule with eleven atoms, and would not be expected to occur in large quantities naturally. Perhaps this quest can be given over to the dedicated advocates of directed panspermia who desire a more suitable home than Earth for the origin of Earthlife.

We do not wish to imply that the macromolecules characteristic of life on Petrolia would necessarily be similar to those on Earth. Rather, the living things would develop along the lines which would function best in that environment and take full advantage of the properties of their native medium. For example, proteins on Earth have polar portions on their outside that are exposed to water, while nonpolar, hydrocarbon-like areas cluster on the inside, avoiding exposure to the water. The organisms on Petrolia might select to reverse the process, using large molecular catalysts that had hydrocarbons on the outside and polar groups on the inside.

The inhabitants of Petrolia might also seek to reverse the direction of certain metabolic processes on Earth. A primary source of chemical energy on Earth is respiration, the combination of organic molecules with the oxygen released in photosynthesis. (See Chapter 3.) On Petrolia it is possible that hydrogen could exist in the atmosphere (if the planet were massive enough to retain it) or be produced by life in a photochemical process. This hydrogen could be combined with suitable "food" to release chemical energy in a reduction process. In this one aspect the inhabitants of Petrolia would be akin to the methanogens of Earth.

Hydrocarbon seas, like ammonia seas, have been suggested as a possibility for Titan, or may exist on another satellite in the outer solar system. If so, we may find out by direct examination what type of life develops in hydrocarbon seas.

3. Silicate Life

The third world that we will visit is the least hospitable of all by our standards. Thermia orbits its sun, one similar to our own, at a distance much less than that of Mercury from our Sun. As a result, its surface temperature is over 1000° C, and permanent seas of lava flow between hills of quartz. Within these seas, silicate beings thrive.

Silicate is a chemical group made of the element silicon, surrounded by a cluster of four oxygen atoms. Earlier we encountered the phosphate group, so vital in Earthlife, which is composed of an atom of phosphorus surrounded by four oxygens. Silicon is in relatively low abundance in the Universe, but it is greatly concentrated in the Earth's crust, where, as silicate, it is a principal component of rocks. The same may well be true in other small planets. Silicon, like carbon, can readily form four bonds and consequently has been the most frequently mentioned alternative to carbon as a basis for life. Advocates of silicon life, however, have usually attempted merely to replace carbon atoms with silicon in molecules typical of Earthlife. Bonds of silicon to itself, or to hydrogen, are weaker than the corresponding carbon bonds. They can decompose on exposure to oxygen or water. The compounds generated by this replacement procedure would be suitable for life primarily at lower temperatures, and in the absence of oxygen and water. Under such conditions silicon might function as an accessory in a carbon-based life form, but unless some unusual fractionation mechanism removed carbon from the environment, we would be unlikely to find life built primarily on silicon.

We come to a different conclusion if we consider not silicon but silicates. (See Fig. 22). Like silicon and carbon, this group has four valences. Silicate units can combine into groups of two, three, or more. They can form chains, rings, ladder-type structures, sheets, and three-dimensional arrays. They can therefore form networks with a complexity equal

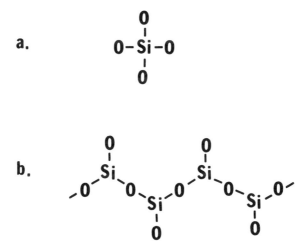

Figure 22. (a) A silicate unit. (b) A chain, made by combining silicate units with elimination of some oxygens. Each silicon atom is still attached to four oxygens, one of which (not shown) is either above or below the plane of the paper.

to that produced by carbon. Aluminum atoms can substitute for silicon in the network to increase the complexity, just as nitrogen does for carbon in our life chemistry. For example, the mineral feldspar consists of a three-dimensional silicate network, with an aluminum atom substituting for every fourth silicon atom. Phosphates and a number of metals can also be present to increase the possibilities for variety. The richness of this chemistry is expressed in the great variety of rocks and minerals on Earth. If a three-dimensional network of silicate units is allowed to extend indefinitely, then we obtain quartz, whose melting point is above 1600° C and is therefore solid, even on Thermia. The analogous network made of carbon atoms is diamond. When the quartz network is interrupted by the regular insertion of metals, then lower-melting crystalline minerals are produced. Common window glass is made of a mixture of silica with oxides of the metals sodium and calcium.

Silicates are not suitable as a basis for life at the temperatures of the Earth's surface, as the stability of their networks keeps them in solid form. Water, the only common liquid on the surface of the Earth, is not a suitable solvent for them. Only at much higher temperatures, sufficient to demolish the carbon compounds that are the basis of our own life, do the silicates soften and enter the fluid state necessary to the process of self-organization.

An energy flow is, of course, also necessary for that process, and on Thermia its own broiling sun supplies that need. Solar radiation absorbed by metals and other impurities in the silicate melt would provide the energy needed to reorganize chemical bonds. Eventually, organized units would evolve that could promote their own replication. The Scottish chemist A. G. Cairns-Smith has suggested that organized mineral surfaces may have played a role in the construction of the first primitive genetic systems on Earth. What they could possibly do for us, they can surely do for themselves! The course of evolution on Thermia could presumably continue to the point where silicate microorganisms, which we will call "lavobes," abound in the molten rock. More complex forms would also evolve, perhaps constructing their shells or other supportive parts out of quartz. Intricate ecologies could eventually arise on such a hot planet.

There is, unfortunately, no planet in our solar system with the necessary surface temperature for such life forms. One other possible type of habitat for lavobes does exist, however, close at hand. The mantle of the Earth exists in partly molten form a few dozen kilometers beneath the surface. It is this reservoir of molten rock that is the source of lava in volcanic eruptions. The molten rock under the crust is called magma rather than lava, and we will refer to the cousins of the lavobes that may live there as "magmobes." These beings could also inhabit the interior of other volcanic worlds, such as Io, a satellite of Jupiter. A problem would be that a suitable form of energy for magmobes would be harder to find inside a planet than on the surface of Thermia.

There are, of course, powerful forces that work within the Earth. They are responsible for abrupt phenomena such as earthquakes and more gradual but awesome effects as, for example, the raising of mountains and drifting of continents. If these processes produced sharp differentials of temperature and pressure within the mantle, unstable chemical species could be generated which would be scavenged for energy by magmobes. The magmobes could, for instance, obtain energy from the formation of stable mineral deposits from less-stable forms. Localized pockets of radioactivity within the mantle are another possible source of heat energy for the metabolism of magmobes. It is even imaginable that magmobes could function as thermophages, gaining energy by moving in cycles between higher and lower temperature levels within the Earth. Since the distance that must be moved to gain useful energy this way is quite a few kilometers, only large magmobes could hope to live this way.

We do not have the means at this time to directly inspect the interior of the Earth for magmobes. It is possible, however, that some may be trapped in solidifying rock beneath the Earth, since creatures of our biosphere are sometimes trapped in mud and sediments. The fossils of magmobes could perhaps then be located in rock samples that have been brought to the surface by geological forces, just as more familiar fossils are found in exposed sediments.

We have described just a few of the many possibilities of alternate chemical life that can exist in the Universe. We will mention others briefly when we consider the environments of particular worlds. Yet others may exist that we cannot imagine because of the limited chemical experience of the human race. When we encounter these alternative forms, we will learn new areas of chemistry by studying them, just as we have learned much of organic chemistry from the study of life forms on Earth.

But first we must search for them and discover them. These events could occur within our lifetimes, when our civilization has gained the capacity to explore other planets in the solar system. This will be done at first by remote orbiters and landers, and, hopefully, later by manned expeditions. The search has already begun. In 1976 two probes of the Viking mission of the U.S. National Aeronautics and Space Administration (NASA) landed on Mars. Their primary objective was to detect extraterrestrial life. In the next chapter we will discuss their results and the prospects for life on Mars.

Chapter 9

Mars

The siren was a beautiful creature in Greek mythology. Her haunting song drove men mad and caused them to steer their ships onto the rocks. The planet Mars has been a siren both for scientific studies and human fantasies about extraterrestrial life. One result of this attraction has been that the words "Martian" and "extraterrestrial" have come to mean the same thing for a large number of people, in and out of science. Ask most people what they think about extraterrestrial life and they will probably answer, "Oh, you mean Martians?" The planet Mars has been the main focus of scientific studies and of speculative literature about extraterrestrial life, at least for the last century.

Although many scientists believe that some of the other bodies in the solar system may prove to be better homes for life, the most elaborate attempt that has been made so far to search for extraterrestrial life was the Viking probe of Mars. Two small vehicles from Earth landed on Mars in the summer of 1976, equipped with ingenious life-detection packages. But in spite of the large amounts of time, effort, and money that went into these probes, their results have been both ambiguous and controversial. They have shed little light on the basic question of whether there is life on Mars, although other parts of the Viking mission did obtain some important information about physical and chemical conditions there. In many respects the Viking life-detection package is a good illustration of how not to search for extraterrestrial life. We will describe the life-detection part of the mission in detail not because we think that the results justify it, but to indicate some of the problems that such searches must overcome.

Since the planning and interpretation of the Viking life search were heavily influenced by a long tradition of dubious and often unconscious assumptions about Mars, we will begin our description with a short summary of the history of speculation about what is or is not present on Mars.

EARLY THOUGHTS ABOUT MARS

For a long time, Mars has occupied the foremost place in the extraterrestrial fantasies of humanity. Moon is a far more prominent object in the night sky; of the planets, Venus is the one which approaches closest to the Earth. Mars, however, unlike Moon, has a measurable atmosphere. Unlike Venus, it affords visible surface features and a temperature range which occasionally overlaps that suitable for living things on Earth. Its reddish color and its name (it is called after a war god) have further attracted interest and stimulated the imagination. Many observers have stared at its hazy markings through telescopes and reported what they wished to see rather than what exists. As early as the 1850's, Sir David Brewster reported that "continents and oceans and green savannas have been observed upon Mars." The start of the modern age of Martian romance came somewhat later, however, in 1877.

In that year the Italian astronomer Giovanni Schiaparelli observed the planet at a very favorable approach and saw linear markings which he called *canali* (Italian for channels). His findings were discounted for a time, but eventually they were supported by others. The noted American astronomer Percival Lowell was particularly enthusiastic. He built an observatory in Flagstaff, Arizona, to study the markings, and also chose to interpret them as canals, the work of intelligent beings. On the basis of extensive observations, detailed maps were prepared (see Plate 8). The features displayed included canals thousands of miles in length, parallel double canals, and elaborate intersections, termed "oases." In the first decade of this century, Lowell publicized a theory that Mars was inhabited by a dying race, fighting nobly for survival on a dried-out planet. The canals were a last-ditch effort to conduct water to arid areas from the polar ice caps of Mars.

The water was needed to sustain the extensive cultivated vegetation of the planet. Seasonal color changes had been observed in the 1880's by a French astronomer, E. L. Trouvelot, and ascribed to plant life. The gradual spread of these color changes from the poles in the Martian spring was said to confirm the effectiveness of the work of the canals. The remaining features of the planet were chosen by Lowell to be compatible with his theory. Extensive canals required the surface of Mars to be very flat. The average temperature was computed to be 9° C (48° F), chilly but bracing.

The theories of Lowell and others stimulated a harvest of derived fiction. In H. G. Wells' famed *War of the Worlds* (1898), tentacled Martian monsters came to Earth looking for an alternative to perishing nobly on their planet. Landing in a suburb of London, they wreaked havoc with heat rays and poison gas. Their efforts to enslave us were ended not by human efforts but by their succumbing to bacterial infection. Forty years later the same story, adapted to radio by Orson Welles and transposed from England to New Jersey, produced considerable panic in the streets of New York. The idea of hostile Martians was vivid and real enough to cause Earthlings to flee their homes.

Other writers, from Edgar Rice Burroughs to Ray Bradbury, have reversed this traffic and sent human beings to

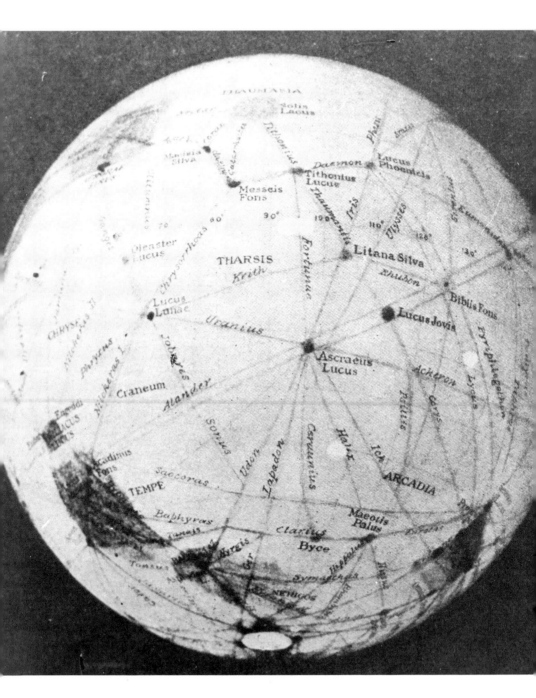

Plate 8.   An early map of Mars, showing canals and oases

Mars. Many wondrous adventures have been set, in the pages of science fiction, amidst the deserts, canals, ruined cities, and vegetation of Mars.

Unfortunately, the factual basis of this romantic Martian fiction was gradually eroded by advances in scientific instrumentation. When better telescopes were constructed, Lowell's canals were resolved into a series of broken lines and other irregular features. Thus perished Martian civilization. In the 1960's and 1970's, the planet was photographed at close range by a series of Mariner spacecraft sent by NASA. In most cases no features at all, not even irregular ones, corresponded with the famed canals. The canals were quite literally in the eyes of the beholders; the results of a strained attempt to resolve objects just at the threshold of vision. This phenomenon has led to other errors in various fields of science.

The vegetation on Mars survived longer in speculation, sustained by other types of scientific measurement. Spectroscopic investigation of radiation from Mars had detected the presence of patterns in the spectrum produced by chemical bonds between carbon and hydrogen. This appeared to confirm the existence of organic matter (presumably the vegetation). In the 1950's, a number of scientific writers argued that the presence of plant life on Mars had been proved beyond dispute. This illusion was again dispelled by the Mariner orbiters. The seasonal color changes were due to dust storms, and the spectroscopic patterns were caused by chemicals in the atmosphere of Earth.

As knowledge about Mars increased, its air grew thinner and thinner, its climate drier and colder. It appeared less and less to be a place suitable for Earthlife. The persistent adherents of life on Mars fell back on the idea of lichens and microorganisms. Of the forms of plant life on Earth, lichens are among the most adaptable to survival under extreme conditions. They are a combination of algae and fungi that live together in a kind of marriage, in which each performs some functions that the other cannot. Lichens hold up well in arid, cold, or exposed environments such as deserts, mountaintops, and the Antarctic. As we have seen, microorganisms are also a very adaptable form of Earthlife. Surely some such species had established themselves on Mars. They could not be detected by telescopes or orbiting space-craft, however. To do that it would be necessary to place a lander on the surface of the planet. Thus, in the early 1960's dedicated individuals began to design life-detection equip-ment for inclusion on such a landing.

The initiative at this time had clearly passed to those skeptical of life on Mars. Speculations in the past had been shown to be fantasy. Why should new ones be entertained? In July 1965, Mariner 4 flew by Mars and photographed 1 to 2 percent of the planet at close range. A landscape rich in craters was observed that in some respects resembled Moon. The skeptics rushed to extrapolate the observations to the entire planet and to picture an aboslutely dead, moon-like Mars. A front page article in *The New York Times* opened as follows: "A heavy, perhaps fatal, blow was de-livered today to the possibility that there is, or once was, life on Mars." The account stressed that there was no evi-dence of river valleys or other forms of water erosion. Mars lacked volcanoes or other mountains. Without the geological instability that produced such features, it would have a monotonously uniform landscape. Mars was both red and dead. Later observations showed this extrapolation to be as fanciful as the inhabited canal world of Lowell. A later NASA report, issued after the Mariner 9 mission, explained the situation:

> *Soon it became apparent that almost all generaliza-tions about Mars derived from Mariners 4, 6 and 7 would have to be modified or abandoned. Participants in earlier flyby missions had been victims of an unfor-tunate happenstance of timing. Each earlier spacecraft had chanced to fly by the most lunar-like parts of the surface, returning pictures of what we now believe to*

be primitive cratered areas. It was almost as if space-craft from some other civilization had flown by Earth and chanced to return pictures only of its oceans.

We now know that Mars has an extensive set of channels and valleys that suggest the earlier presence of liquid on the surface. The volcanoes on Mars tower to heights far above those on Earth. The Martian terrain is in fact quite varied and unlike that of the Moon. The psychological need in some human beings to deny the existence of extraterrestrial life is as powerful as that of others to assert that it must exist. In either case the result is the same. There is a temptation to rush far beyond the evidence (and away from reality) in order to reach the much desired conclusion.

THE VIKING MISSION

The initial series of close scientific observations of Mars culminated with the Viking project, which placed duplicate automated laboratories on the surface during the summer of 1976. This feat marked the climax of an eight-year effort involving thousands of individuals. Two 3,500-kilogram spacecraft had been launched in 1976. (Plate 9) They completed a ten-month voyage of 710 million miles by assuming an orbit around Mars. Mars is not that far from Earth, but a spaceship must follow a complicated curved path to get from here to there, and so must travel much longer than the straight-line distance. The orbiter component of the spacecraft conducted a photographic reconaissance to select suitable surface locations for the automated laboratories. This was necessary because of the design of the 600-kilogram lander (Plate 10). The instrument package was supported by three legs, and would have been damaged if any leg had chanced to land on a boulder larger than twenty-two centimeters.

There was considerable debate over the virtues of alternative landing sites, and much anxiety during the landing

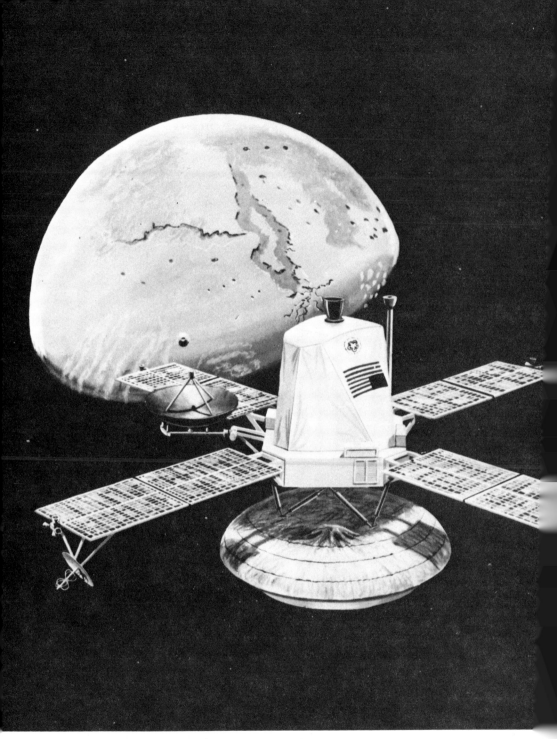

Plate 9.   A simulation of one of the Viking spacecraft, in the vicinity of Mars

Plate 10. A Viking lander as it might appear to a hypothetical inhabitant of the surface of Mars

process. Since Mars is far from Earth, it takes several minutes for our radio messages to reach it. An error in the terminal stage of landing could not have been corrected by instructions from Earth before damage occurred. During the landing of Viking 2, communication with it was lost for a time, and there was fear for its safety. In the end all turned out well, and both spacecraft landed safely. One landing

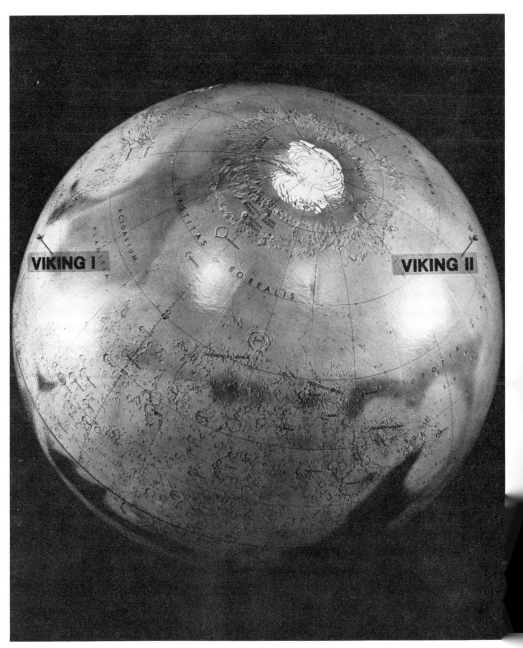

Plate 11.   The globe of Mars, showing the two Viking lander sites. It may be compared with the earlier version by Percival Lowell (Plate 8).

site was in an area called Chryse, at 22.5° N latitude. The other was on the opposite side of the planet, at Utopia, 48° N latitude (Plate 11). The former site was the warmer one, while the latter was thought to contain more moisture. It had been expected that the terrain around one Viking site would differ from that of the other.

The pictures transmitted by the television cameras mounted on the two landers were similar, however. They showed flat to slightly rolling deserts, colored a mixture of red, brown, and yellow and strewn with rocks and boulders, beneath an orange-yellow sky (Plates 12, 13). The sky color is the result of dust in the Martian atmosphere, spread by the incessant winds. In subsequent months of observation,

Plate 12. The first photograph taken from the surface of Mars, showing the area near the Viking 1 lander. A footpad of the lander is on the right. The rock on the left is about 10 cm. across.

Plate 13. A Martian landscape, taken by the Viking 2 lander, showing rocks, soil, and a very thin coating of what is thought to be water ice.

neither camera revealed any moving object or other familiar sign of life. The burden of testing for life, and of studying the chemistry of the surface, fell on the complex instruments, which had been delivered safely. They furnished data for several years and added much to our understanding of Mars, as well as confusion about whether life exists there.

The results of the Viking project, taken together with

earlier observations, have given us a more detailed idea of the physical nature of the planet than we have ever had before. We still have explored only a very small part of its surface, however, and many gaps in our knowledge remain to be filled. It is likely that future exploration of Mars will change our views about it as much as each previous mission has.

THE CHEMICAL AND PHYSICAL NATURE OF MARS

A picture of the present situation on Mars and its past history has emerged, based on the best information that we have at this time.

The radius of the planet is about three thousand kilometers, roughly one-half that of Earth. As there are no Martian oceans, the land area is about the same on both planets. Mars is one and a half times farther from the Sun than it is from the Earth, and its orbit is more elliptical. Less than half as much sunlight reaches each point on Mars. This is partly compensated by the fact that the Martian atmosphere absorbs less of the light except during dust storms.

Mars rotates on its axis about once in an Earth day, and it takes almost two Earth years to revolve about the Sun. The axis is tilted with respect to its orbit, as is that of Earth. This, combined with the fact that the orbit of Mars is more elliptical than that of Earth, leads to seasons that are more pronounced than on our planet. The overall result is that Mars is colder than Earth. Over the whole planet and year, the maximum temperature range is from $-140\,^{\circ}$C to $+20\,^{\circ}$C. At the Viking lander sites, the temperature varies from $-90\,^{\circ}$C at night to $-10\,^{\circ}$C during the day. So, for most of the planet, most of the time, the temperature is below the freezing point of water.

The orbiting spacecraft took pictures of Mars that could detect an object about as small as a football field. These pictures reveal many remarkable and varied surface features both similar to, and different from, those on Earth. The most striking are a large number of huge craters, especially in the southern hemisphere. These craters are similar to those found on Moon and are the result of countless meteorites that hit Mars, mostly in the early days of the solar system, four billion years ago. The fact that they have persisted without distortion for so long is an indication that much less weathering and moving of the surface has taken place on Mars than on Earth. But in the northern hemisphere there are many fewer craters, and the surface is dominated by smooth plains of volcanic lava, strewn with boulders. Two sites of this type were sampled in detailed by the Viking landers. Some lava flows extend for kilometers, and at least some were formed hundreds of millions of years ago.

The northern hemisphere is also marked by several enormous volcanoes. The largest, Olympus Mons, is twenty-three kilometers high, far higher than any mountain on Earth, volcanic or not. It is thought that these great heights are possible on Mars because there is less motion on the crust than on Earth. Olympus Mons was formed perhaps two hundred million years ago, while other volcanoes date back to the time of formation of the planet. It is possible that active volcanoes still function, but no observations supporting their existence have been made up to this time.

No bodies of liquid water have been observed on the surface of Mars, nor would any be expected under the existing conditions of low temperature and low pressure. However, the pictures taken from the orbiters show a series of "rills," which look to all intents and purposes like rivers and their tributaries (Plate 14). The age of these channels has been estimated to be between one-half billion and three billion years, although exact dating is impossible. This discovery raises the important possibility that liquid water did exist once on the surface of Mars. Further evidence for this came from the observation that in many places, the terrain has

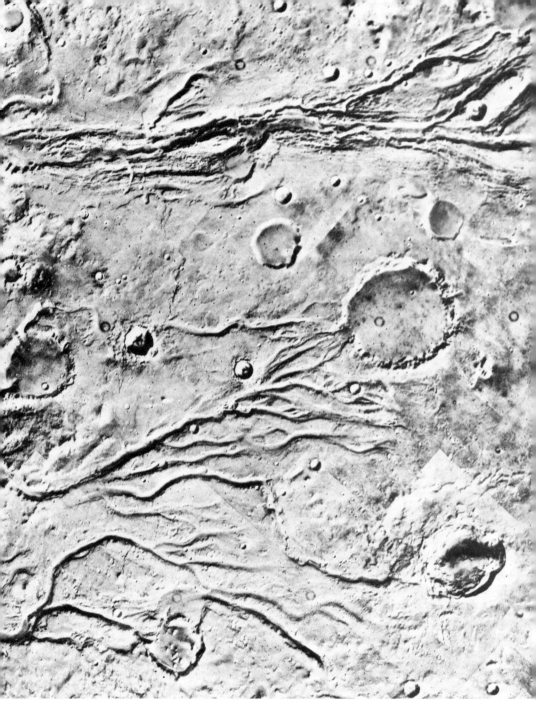

Plate 14. Martian channels, photographed by the Viking 1 Orbiter. Each channel is many miles long. It is thought that they were cut by an ancient flood, in a period when liquid water could exist on the surface of Mars.

collapsed in a way that is suggestive of ground ice (permafrost) which has been suddenly heated by volcanic action, and then melted. If this happened now, the water would evaporate very quickly and would not form the long channels that are seen. But at an earlier stage in Mars' history, if the overall temperature and air pressure were higher, the liquid water could persist long enough to form rivers. If this was the case, several questions arise. For how long did bodies of liquid water exist on the surface of Mars? Where did the permafrost originate, where did the water go after the rivers dried up, and why was the temperature on Mars much higher in the past than it is now? Only suggestions can be given as answers to each of these queries, and new information about Mars is needed to confirm these suggestions or to provide better ones.

Other surface features relevant to the water question are the Martian polar caps originally seen from Earth. Measurements of temperature and water vapor over these caps have demonstrated that for the northern polar cap there is a permanent core of water ice mixed with dust, covering a few percent of the planet's area and upward of one meter thick. There are also temporary ice caps found in winter that apparently consist of solid carbon dioxide (dry ice), which condenses out of the atmosphere at low temperatures. It is possible that the permanent ice caps contain most of the water that once ran liquid in Mars' rivers. Alternatively, that water may now exist as permafrost spread over much of the planet. In either case the total amount of liquid water on Mars was probably much less than that present on Earth, perhaps corresponding to a number of shallow seas in various parts of the planet, rather than to Earth's oceans.

The orbiter and Viking lander also measured the composition of Mars' atmosphere, which is quite thin, only about 1 percent as dense as Earth's at the respective surfaces. Mars' atmosphere has a very different composition than that of Earth, being 95 percent carbon dioxide, with small amounts of nitrogen and argon and almost no oxygen.

This atmosphere is perhaps similar in composition to the one Earth had long ago before photosynthesis and other bacterial action liberated the large amounts of oxygen now present. But this does not indicate that photosynthesis has not taken place on Mars, since certain types of photosynthesis are known which do not release oxygen. In addition, the type of photosynthesis which does release oxygen took place on Earth for many millions of years before a substantial buildup of this gas in the atmosphere occurred.

The lack of oxygen on Mars does result in there being no ozone in its atmosphere. There is nothing to keep much of the ultraviolet light from the Sun from reaching the surface as ozone does on Earth. This strong ultraviolet light may have an important effect on the surface chemistry and possible biology of Mars.

There is water vapor in Mars' atmosphere, again in much smaller quantities than on Earth. The region of maximum water vapor, over the polar ice cap, is as dry as the driest parts of Earth, in Antarctica. There is only 1 percent as much water vapor in the atmosphere here as at a typical site on Earth. The average place on Mars, such as the Viking lander locations, is even ten times drier. If there is still volcanic activity on Mars which evaporates some of the permafrost, then this may lead to short-term increases in the water vapor content of the atmosphere, and it would be interesting to see if these can be detected. Liquid water might exist in underground caves on Mars, where the temperature is high enough to melt the permafrost. It also might exist as ground water some distance below the surface. In two areas south of the Martian equator, the presence of unusual concentrations of water vapor has suggested to some scientists that ice may exist just inches under the surface sediment. These scientists believe that under certain circumstances, this ice could give rise to liquid water, which would exist temporarily at or near the surface. Discovery of such bodies of water awaits much more detailed exploration of Mars, probably with mobile landers.

Direct analysis of the surface material at the Viking lander sites showed it to be mostly a mixture of many of the same kinds of minerals found on Earth, primarily iron and silicon oxides. Some water was present, bound to the minerals. The yellowish red-brown color is apparently caused by the presence of an iron mineral called limonite, which has such water bound to it. Perhaps some of the liquid water that once existed ended up in the limonite.

Although the bulk composition of the minerals on the surface of Mars was mundane, their chemical properties were quite exotic, unlike those encountered in mineral samples from Earth, Moon, or falling meteorites. These properties were discovered during the life-detection experiments and will be discussed in detail shortly. For now, we can mention that the most spectacular property of the soil was its ability to react with added water, releasing oxygen. If this were due to a chemical reaction, then Martian soil contains a chemical that is a better oxidizing agent than oxygen itself. Such chemicals have been produced in laboratories, but they have not been observed in mineral samples taken on Earth from meteorites, or on Moon. One known class of such chemicals is called "superoxides," and this term has also been applied to the unknown substances on the surface of Mars. We would not expect to encounter such chemicals on Earth because the large amounts of water present would tend to destroy them. It is less clear why the oxygen-releasing chemicals were absent on Moon. The properties of known superoxides are somewhat different from the properties of Mars soil, and no direct evidence for superoxides has emerged from chemical analyses of the surface. One proposal has been published that attributes the release of oxygen to a physical effect rather than to a chemical reaction. The oxygen may have been released from tiny pores in the soil by the added water. Such speculations are fascinating, but we will have to wait for the results of the next Martian lander before we can have a definite answer.

One type of chemical analysis was thought to be of particular importance in the search for life on Mars: the measurement of the amount and kinds of organic compounds present. As we have seen, carbon compounds make up much of the bulk of living creatures on Earth and play a central role in their metabolism. Samples of Earth soils, even from deserts, generally contain a wide variety of carbon compounds. They are derived from the debris of dead organisms, which is often present in thousandfold greater amounts than the living organisms themselves.

Because of the wide distribution of living creatures on Earth, it is actually difficult to locate soils free of the organic compounds derived from life. The most barren samples come from the dry, cold Antarctic valleys that we discussed earlier. Certain, but not all, samples from these areas were free of living organisms and of all but trace amounts of organic compounds.

It was expected that living creatures on Mars would be made of carbon compounds. As stated before, we believe that living beings may be constructed in ways that do not use carbon compounds in vital roles. However, carbon is a versatile element and is readily available, as carbon dioxide and carbon monoxide, in the Martian atmosphere. The temperature range on Mars, while perhaps too low for some of the specific reactions of Earthlife, is still suitable for other organic reactions. Mars is sufficiently similar to Earth in composition and environmental conditions that life based on carbon chemistry is a reasonable possibility there.

The presence of organic compounds in Martian soil, of course, would not necessarily have indicated that living creatures were there. Certain meteorites, the carbonaceous chondrites, contain organic compounds that are produced by nonbiological processes. It is likely that a considerable number of them fell on Mars during its history, distributing carbon compounds on the surface. Model experiments had also shown that the ultraviolet part of sunlight could cause the synthesis of organic compounds in an atmosphere like that of Mars. It was thought that some of the compounds would "rain" onto the surface and remain there.

To distinguish among these possibilities, a detailed analysis of the identities and amounts of any organic compounds in the Martian soil was necessary. From this information it was hoped that the various alternatives could be sorted out: meteorites, nonbiological synthesis in the atmosphere, life resembling Earthlife, or a pattern that resembled none of the above but perhaps signified the presence of a new type of life. The constraints on the Viking mission permitted the use of only a lightweight, automated instrument for organic analysis. One device seemed particularly suitable: a combination of a gas chromatograph (GC) and a mass spectrometer (MS).

The GC is useful for separating a mixture of gases into individual components. For example, on Earth it has been used to study coffee aromas and perfumes. The GC is not effective in directly studying a solid, such as a raw steak. On broiling a steak, however, various components break down to release its characteristic aroma. This mixture could be analyzed by GC, and an experienced chemist could use the information to deduce that it was a steak, rather than some other food, that had been cooked.

If an entirely new substance such as Martian soil, instead of a familiar one like a food, were being heated, a GC alone would not be enough for a full analysis. The GC would tell you that gases were being released, and how many components were present, but not the full identity of each one because it would not be known in advance what substances might be present on Mars. The final identification was the function of the MS. A combined GC-MS instrument had been used in laboratories on Earth to study samples of meteorites, typical soils, and even pure bacteria, and it functioned well. The patterns obtained from Earth soils resembled those produced by bacteria, but differed markedly from the meteorite analyses. Two duplicate lightweight GC-MS instruments were safely deposited on the surface of Mars by the Viking landers, and the organic-analysis data was awaited eagerly by teams of trained investigators.

When the results came, they were astonishing. No carbon compounds of Martian origin were detected, within the delicate sensitivity of the instrument, at either Viking lander site. There was no data to be analyzed, no pattern to interpret. Some known organic impurities, introduced into the GC-MS during its manufacture, were detected. This indicated that the instrument was working well. But nothing else was observed. The level of organic compounds in the soil of Mars was less even than the bare traces that had been seen in lunar samples.

This result, while starkly negative, was still subject to alternative explanations. It was possible that no carbon-based life was present on Mars, and that the amount of organic compounds produced by meteorite infall and synthesis in the atmosphere was considerably less than expected. It was also conceivable that organic compounds were being produced by one or more of the above processes, but that they were then destroyed on the surface. A number of agents have been proposed that could destroy organic compounds, such as superoxides, or the combination of oxygen, ultraviolet light, and a catalyst from the soil. The possibility also existed that Martian microorganisms, unlike Earthlife, actively scavenge and use their own debris, as well as any organic material that rains down. In that case the GC-MS could detect life only by responding to the organisms themselves. It has been estimated that as many as one million Earth bacteria could be present in a gram of soil and still fall below the sensitivity of the GC-MS. This would not be a remarkably low population, even by Earth standards. Several typical terrestrial soils used to calibrate one of the Viking life-detection experiments had microorganism counts in the range of two to four million per gram.

In summary, the GC-MS analysis did not furnish the harvest of data that had been expected. It demonstrated that the Martian soil at the two sites analyzed, unlike soils on Earth, is extremely poor in organic compounds. No evidence supporting the presence of life had been obtained, but, in the words of the investigators themselves, "The demonstration that very little, if any, organic material is present does not exclude the existence of living organisms in the samples analyzed."

THE LIFE-DETECTION PACKAGE

The organic analysis instrument, while valuable in itself, has never been intended to bear the main burden in identifying the presence of life on Mars. That role was filled by a separate set of experiments, especially designed for biological purposes.

The scientific community was specifically invited by NASA in 1969 to suggest suitable experiments for the detection of life on Mars. For a number of years before that time, however, there had been a strong consensus within NASA that the search for extraterrestrial life, particularly on Mars, should be a prime goal of the unmanned space program. A number of prominent scientists had concerned themselves with the difficult question: How does one design an instrument for the remote detection of life? This is a particularly challenging problem when the nature of both the life form and the surrounding terrain are unknown. Several of the experiments that were included on the Viking mission were originally designed in the early 1960's. At that time, little detailed knowledge was available about conditions on the surface of the planet. Vegetation on Mars was still considered to be plausible, and canals were not fully eliminated from consideration.

Apart from the scientific problems involved in the Viking life-detection experiments, there were additional difficulties posed by the economic circumstances of the project. Financial and safety considerations severely limited the amount of payload to be delivered to the surface. Furthermore, Viking was designed on a one-shot basis. In the Apollo project, a series of Moon exploration voyages had been planned, so that discoveries on one mission could be exploited on the next. In the case of Viking, very little thought was given to a follow-up to the first search, and no commitment was made. Those circumstances encouraged the design of "jackpot" experiments—ones that could provide convincing evidence for the presence of life if they were positive. Such experiments had to make a number of assumptions based on our experience with Earthlife.

Seventeen proposals for life detection were submitted in response to NASA's 1969 request. Four were selected for the Viking biology package. One was later cancelled due to time and money pressures, but three eventually landed. By the end of the selection process, a general philosophy had emerged:

1. The tests would aim at detecting microorganisms. On Earth, microorganisms show the ability to survive in extremely varied environments. They occur whenever larger organisms do, while the opposite is not always the case.

2. A variety of experiments would be selected. Each would be based on different assumptions, measure different features, and be analyzed separately. In the words of Richard Young of NASA, "It was felt that any one technique for life detection was unlikely to succeed, and therefore, a combination of techniques was required." An individual experiment might be based on incorrect assumptions about the nature of Martian life and so miss it, although life was present. The same unfortunate result could also occur with the final package, but at least Martian life would get three strikes before it was ruled out.

3. The experiments that were selected all tested for the growth or metabolism of microorganisms. Other concepts had been considered. The inclusion of a microscope would have permitted a search for the shape or motion of microorganisms. As we discussed earlier, many chemicals vital to Earthlife are employed in only one of their two possible mirror-image forms. An instrument could have been included

on the Viking probe to test whether unequal amounts of mirror-image forms of chemicals existed on Mars. These and other ideas were sound in concept, but were eliminated for technical reasons, such as insufficient sensitivity.

4. Two of the three life-detection experiments that flew on the Viking mission attempted to culture Martian organisms under conditions more typical of Earth than of Mars. Samples to be tested were added to water solutions of chemicals known to be nutrients for Earthlife. The testing chamber had to be heated to 10°C to keep the water in liquid form. There was an obvious risk that indigenous life might be killed by the conditions used. It could be poisoned by one of the nutrients, or killed by the high temperature. Liquid water is rare or nonexistent on Mars, and a local organism might respond to it as we would to a vat of boiling oil.

Let us use a diagram to express the logical possibilities involved in a test for life on Mars (Fig. 23).

Positive or negative results are those that indicate the presence or absence of life, respectively, according to criteria set in advance of the Viking mission by the designers of the

EXPERIMENTAL RESULTS	ACTUAL SAMPLE	
	Life	*No Life*
POSITIVE	1	2
NEGATIVE	3	4

Figure 23. The logical possibilities involved in a test for life on Mars

experiments. Box 1 represents the case where existing life is successfully detected, while box 4 stands for the situation in which life neither exists nor is detected. The other boxes represent misleading results. A false negative would be obtained if existing life were missed by the experiments (box 3). A false positive would be registered when nonliving materials afforded a lifelike result (box 2).

Because of the design of the life-detection experiments, the various possibilities were not equally likely, as Fig. 23 suggests. This situation can be emphasized by redrawing the diagram so that the sizes of the boxes reflect the altered expectation (Fig. 24).

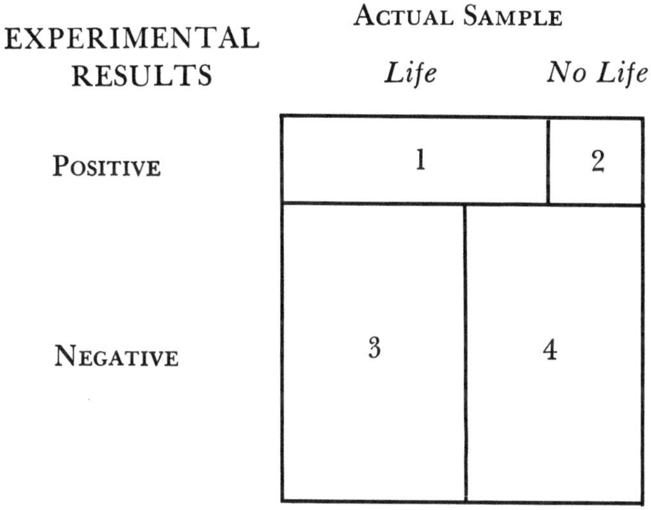

EXPERIMENTAL RESULTS | ACTUAL SAMPLE

	Life	*No Life*
POSITIVE	1	2
NEGATIVE	3	4

Figure 24. The logical possibilities involved in a test for life on Mars, as affected by the design of the Viking experiments

The use of specific tests, based on assumptions derived from Earthlife, had maximized the chance that a positive result would be significant (box 1), rather than an artifact (box 2). Furthermore, each experiment had been tested for years in order to eliminate any possible false positive results. The Earthlife-oriented design of the tests, however, made a negative result more likely than a positive one, with a substantial chance that a false negative (box 3) could occur. In the words of the astronomer Carl Sagan and the Nobel Laureate microbiologist Joshua Lederberg, "Negative results would exclude an important subset, but only a subset of the possible classes of Martian organisms."

The results of the Viking biology package were, on the whole, positive. Options 3 and 4 listed above were eliminated. A continuing debate has arisen between the advocates of possibilities 1 and 2: life on Mars versus a set of chemical and physical effects that imitate the responses expected of life. Mankind's first ambitious attempt to detect extraterrestrial life at close range has resulted in ambiguity and controversy. This situation is fully in accord with the past tradition of Mars observation.

In order to understand the nature of the Viking life-detection results more fully, we have interviewed the scientists in charge of each of the three experiments, as well as NASA officials responsible for coordinating the results of the entire package. The next sections reflect our impressions and interpretation of the published results as modified by these interviews.

THE GAS EXCHANGE EXPERIMENT (GEX)

Of the three life-detection experiments that went to Mars, only the gas exchange experiment (GEX) originated within NASA. Its designer was Vance Oyama, a Japanese-American biochemist with considerable experience and expertise in the design of instruments. He had had responsibility for the detection of possible microorganisms in the soil samples brought back from Moon.

Vance Oyama's laboratory is located at NASA's Ames Research Center, a huge collection of aircraft hangars, Quonset huts, and squat concrete buildings near the south end of the San Francisco Bay area. When interviewed at his lab, Oyama emphasized the various difficulties that had to be overcome in order for his experiment to arrive and function on Mars.

The GEX test had come close to being eliminated in the selection procedure. A number of potential instrumental malfunctions subsequently had to be detected and corrected within a short period of time. In the end Oyama was quite satisfied with the performance of the apparatus.

To understand the idea of the GEX experiment, consider the fate of a bowl of clear beef broth left standing on your kitchen table. After a few days, your nose will tell you that something has happened to it. Microorganisms have spoiled the broth. In the GEX test a mass spectrometer (MS) takes the place of your nose and tells you of changes that occur in the gases above the broth. In the actual design, a suitable broth was added to a sample of Martian soil in a sealed vessel under a simulated Martian atmosphere, but at an Earthlike temperature (required to keep the water liquid). The atmosphere contained carbon dioxide and inert gases but lacked oxygen. In tests with Earthlife, only those soil organisms that could grow without oxygen (were anaerobic) responded. Fortunately, all Earth soils tested that contained viable organisms also had some anaerobic species. Complex changes took place in the levels of the various gases that were observed. However, the types of response varied; there was no clearly defined criterion for a positive result. In interpreting the patterns, one had to rely upon past experience with a variety of samples.

Unfortunately, there was some danger of a false positive result. Sterile soils could on occasion release gases by chemical reaction. Some gases could be held mechanically on the soil by a physical process called adsorption, and be released when a new solution was added. Two options were available to distinguish these results from the responses of microorganisms. The soil could be preheated at 145°C before the test. Living organisms are usually killed at this temperature. A response to the test that was eliminated when the sample was preheated would presumably be due to life. An alternative procedure was to remove the broth and gases after a positive response, and add fresh ones. A growing colony of organisms would greet each new meal with the same enthusiasm that it had the first, while a chemical response would eventually be exhausted by successive cycles.

The broth used, of course, was not beef broth, yet it was one that any doting mother on Earth could serve proudly. It contained over fifty ingredients, including amino acids, vitamins, minerals, and other substances, each of which had been shown to be essential to the growth of at least one Earth organism. Some possible ingredients had been eliminated and others kept at low concentrations, to avoid levels demonstrated to be toxic to some microorganisms. As a concession to the hypothetical Martians, the mirror-image forms of the amino acids were included in addition to the natural ones. However, if the Martians also needed the mirror-image forms of the vitamins for growth, they were out of luck, as these were not supplied.

The use of this complex growth medium created ample opportunity for false negatives. We have already discussed the possibility that the local organisms might be poisoned, drowned, or boiled by the medium. If they escaped these fates, they might still lack some essential Martian growth factor. A positive response would also be missed if it were too slow. One sample on Earth, containing a small number of microorganisms, had responded only after two hundred days. The longest time allowed on Mars was 201 days. Very little margin had been left for interplanetary variability among species!

Because of these considerations and in order to simplify the data, a modification was introduced into the GEX experiment before it was launched. In the first step, the soil sample was exposed to water vapor alone. This step allowed the effects of exposure to water to be separated from those produced by immersion in the complex broth. The late Dr. Wolf Vishniac (a biologist at the University of Rochester, New York) had demonstrated that the addition of water alone to a soil sample was sufficient to allow the growth of microorganisms. The other necessary ingredients for growth were leached from the soil by the water. In practice, all the important data of the GEX experiment was produced by the initial humidification with water vapor, and none in the main body of the experiment.

When Martian soil from the Chryse site was humidified with water vapor, an unexpected result took place: Oxygen was released. This spectacular result was accompanied by the release of considerable amounts of carbon dioxide and nitrogen. The types of chemicals which are capable of liberating oxygen from water, such as the already mentioned super-oxides or corrosive fluorine gas, had hitherto been encountered only in laboratories and not on planetary surfaces.

Similar results were obtained with samples taken at the second Viking site, Utopia. One sample was taken from under a rock, to test whether the exposure of the soil to the sunlight was necessary for the oxygen release. Considerable practice and some ingenuity were needed in order to use the Viking-lander sample arm to move a rock aside. In the end the result was the same; samples taken from the shade and in sunlight both released oxygen upon humidification.

A sample of soil from each Viking site was preheated at 145°C and then tested. In one sample, half of the usual amount of oxygen was released, while in the other, no oxygen was released. The discrepancy was ascribed by Oyama to instrument malfunction. The release of carbon dioxide was abolished in the two preheated samples, while the nitrogen release was relatively unaffected. The results were thus ambiguous about the effect of heating the samples. It was unfortunate that circumstances did not permit the performance of additional runs, to clarify the situation with respect to oxygen release after heating.

If the results obtained upon merely humidifying the Martian soil were surprising, contrary to all previous experience, and not readily explained, then little clarity could be expected from the main part of the experiment, the addition of the nutrient broth. A number of gas changes were observed, and the patterns did not resemble any previous tests with other soils. Some changes persisted through several cycles. It is difficult to draw any firm conclusion from these results, since any additional effect of the nutrients was likely to be swamped by the large effect of the water alone.

How are the results of the GEX experiment to be interpreted? Do they support or deny the existence of life? We feel that the wisest course is not to draw any conclusions at all. In its applications on Earth, the interpretation of GEX results had relied heavily upon prior experience with well-characterized life-bearing and sterilized soils. The results of the humidification step immediately established that Martian soil was unlike all others that had been tested. Under these circumstances, no sound conclusions can be drawn about the causes of the unusual effects that were observed.

It is, of course, useful to construct models of what might have occurred. A number of competing proposals have been presented that advocate nonbiological explanations. In some cases the proposals have been accompanied by simulation of a portion of the laboratory results. Gas changes have been attributed by some to physical adsorption and release processes. Chemical changes due to superoxides have been invoked. In some cases solar radiation has been given a role (although oxygen release was also observed with the sample taken under a rock).

One of the most speculative explanations of the Viking results has been devised by Vance Oyama himself. He proposes that chemical evolution has taken place in a dry fashion in the Martian atmosphere, to produce complex polymeric molecules. These molecules are now present on the surface. They are responsible for the observed color changes of the surface and many of the Viking results. They elude detection by the GC-MS for technical reasons. With the addition of moisture and some further evolution, the molecules could give rise to DNA and other precursors of life. In our interview Oyama expressed confidence that some of the steps could be duplicated in the laboratory. We feel that this scenario is quite unlikely. It brings together a large number of hypotheses which have not been supported by independent experiments. In addition, there are intrinsic difficulties involved in conducting chemical evolution without a suitable medium to promote reaction rates, as was discussed in Chapter 7.

Perhaps one of these chemical explanations may prove true. It is also possible that a novel life form, more heat resistant than ones on Earth, is responsible for the results. These speculations are useful in that they help in the design of future experiments. But we must remember that they have been invented after the fact to explain the unexpected data. None of the explanations can be accepted until its predictions have been confirmed by additional experiments on Martian surface material. The GEX results, in their present state, tell us little about the presence or absence of life on Mars.

LABELED RELEASE (LR)

Dr. Gilbert Levin's laboratory is located in a modest commercial building in a suburban industrial lot in Maryland. A tall, friendly, enthusiastic biologist, Dr. Levin is the president of Biospherics Inc., a company that consults for the government on problems involving microorganisms. He was quite happy to show us around the building, particularly the laboratory which housed the earlier, homemade version of his life-detection apparatus, nicknamed "Gulliver." Gulliver had been in the design stage for years before the official invitation to participate in the 1969 Viking project. After selection, this experiment went by the more austere name of labeled release (LR).

The setting where the LR experiment was designed was in striking contrast to the sprawling NASA location in California that gave birth to the GEX. Nevertheless, the two experiments were closely related in concept. Both were attempts to cultivate Martian organisms in a nutrient broth nourishing to Earthlife, and to detect them *via* their growth and metabolism. Also, both the LR and GEX experiments were more sensitive to a type of microorganism that consumes organic chemicals produced by other sources, either living or nonliving, as opposed to blue-green algae, which use sunlight, water, and air to produce their own organic chemicals.

Dr. Levin's broth was simpler, the question asked more specific, and the answer more readily interpreted. Only five simple chemicals were supplied. Where mirror-image forms of a chemical could exist, both were given. The carbon atoms in these chemicals were tagged with radioactivity. Their fate could be followed, regardless of whatever else happened within the experimental vessel. The use of radioactive compounds also made this experiment much more sensitive than the GEX, as radioactivity counters can detect much smaller quantities of material than can a mass spectrometer.

The key question asked in the LR experiment was whether the radioactive substances were being converted to a gas. A portion of the food that we, and other forms of Earthlife, eat is oxidized in our bodies to carbon dioxide and released in our breath. If Martian microorganisms were similarly to "eat" any of the chemicals supplied and release the carbon atoms in a gas, this would be recognized by the radioactivity detector.

The chances for a false negative were similar to those in the Oyama experiment. The Martian organisms might not survive in a medium and at a high temperature far beyond what they ordinarily experience. If they did survive, they might be poisoned by the chemicals offered, or at least not choose to eat them. There was also the chance that they might use them but not convert them to a gas. If so, no growth would be detected. A negative test would not be conclusive evidence against the presence of life.

The safeguards against a false positive result resembled those in Oyama's gas exchange experiment. One could sterilize the soil by preheating it at 160°C. If the response were due to life similar to Earthlife, it would be eliminated by preheating. After a positive release of gas, a fresh supply of broth could be added. A chemical response would presumably eventually diminish, while a response due to life would continue to occur. In extensive tests with Earth soils, fertile samples uniformly gave positive responses, while barren ones were negative; no radioactive gas was released.

The LR experiment gave a clearly positive result for life on Mars. Samples from both Viking sites were tested, including the one scooped from under a rock, with very similar results; radioactive gas was released. The amount of gas released suggested that only a small portion of the chemicals offered were being used, perhaps only the simplest one, a compound called formate. The experimental design did not allow this point to be tested directly. Dr. Levin offered an explanation in our interview. He felt that additional chemicals had been consumed, but that a portion of the carbon dioxide that was generated was being retained on the soil and in solution.

In fact, when additional broth was added to initiate the second cycle, there was a rapid uptake by soil or solution of a portion of the radioactive carbon dioxide that had already been released. Similar effects had already been observed in the GEX. These effects may have been caused by a physico-chemical process irrelevant to the search for life. Unfortunately, this result made it difficult to determine whether the chemicals in the second cycle of nutrient broth were being consumed. After the reabsorption of gas, the release of the gas resumed at a slow rate. Was this just a reversal of the absorption, or was additional "food" being consumed? To obtain more information, the broth was warmed at the end of a second cycle. A quick burst of gas release was observed. A large amount of gas had been retained in the broth and had not been detected previously. Quite possibly a portion of this gas had been generated in the second cycle.

The crucial test for life was the preheating step. When a

sample was heated at 160°C and then tested, no release of radioactive gas was observed. Because of the great importance of this result, a much more rigorous test was introduced. A method was found to adapt the heater so that heating could be carried out at about 50°C rather than at 160°C (this need had not been anticipated in the original design). It was thought quite unlikely that a chemical process would be impaired by this temperature. Living organisms, particularly cold-adapted Martian ones, might be affected. In fact, much of the ability of the soil to release radioactive gas was destroyed by the 50°C treatment. It was later noted that simple storage of the Martian soil within the instrument (at 10°C) for several months was sufficient to inactivate the agent causing the release of radioactive gas.

The results correlate quite well with the idea of sensitive microorganisms that start to die when introduced to the experimental conditions, but survive sufficiently long to produce a clear positive result. In the words of Harold Klein, the NASA biologist who coordinated the three biological life-detector experiments, "In many respects, the LR data was entirely consistent with a biological interpretation. Indeed, if information from other experiments on board the Viking landers had not been available, this set of data would almost certainly have been interpreted as presumptive evidence for biology." We must keep in mind, of course, that each experiment *was* intended to be a separate probe for life. It was felt possible, even probable, that one test might succeed, while others, because of limits in their assumptions, would record false negatives.

Efforts have been made to construct alternative chemical explanations for the LR results. It is an everyday fact of life in chemical laboratories that a wide variety of chemicals can convert organic substances by oxidation into carbon dioxide. Of the LR chemicals, formate is the most readily oxidized, but given a sufficiently strong oxidizer, almost anything will be attacked. A favorite in our laboratory is a hot mixture of chromic and sulfuric acids, which we use as a cleaning solution. It will, in a short time, clean out almost any tarry, intractible, organic mess. We have to be careful because it will also happily attack the floor linoleum, our clothing, and our fingers. It is thus an easy matter to suggest chemical reagents that could convert the LR broth ingredients into carbon dioxide. In advance of Viking, it was considered unlikely that such chemicals would be present in soil samples. The LR experiment had successfully distinguished soils containing living organisms from sterile soils in all tests conducted on Earth. It further stretches the imagination to believe that such chemicals would be inactivated at precisely the temperature that suffices to kill living organisms.

The experiment was not designed, of course, to be decisive in itself about life on Mars. In terms of the actual situation, in which the GEX and other experiments did show that the Martian soil has peculiar and unexpected properties, we must be cautious in drawing a firm conclusion that the LR result establishes the presence of life on Mars. Nonbiological explanations are speculative but possible. They cannot be promoted from speculations to accepted facts, however, just because it is convenient to do so. If we choose to disregard or circumvent the clear results of life-detection experiments, then it becomes pointless to send such experiments at all.

THE CARBON ASSIMILATION EXPERIMENT

This was the only one of the life-detection experiments that was designed to permit success on Mars but fail on Earth. The microorganisms on our planet, when tested by the apparatus, did not respond to the dry form of this test, conducted in the following manner.

A sample of Earth soil was exposed to an atmosphere of radioactive carbon dioxide and carbon monoxide. The latter gas, while poisonous to human beings, is tolerated and used by a number of microorganisms. It was included because it is a minor constituent of the Martian atmosphere. The other gases of the atmosphere of Mars were also present, but moisture was absent. Simulated Martian sunlight (emitted from a special lamp) shone brightly. After a time, the soil was analyzed to see if any of the radioactive gases had been taken up and converted to organic compounds. This had not occurred.

In concept, the carbon assimilation experiment (sometimes confusingly called pyrolytic release or PR) was the reverse of Levin's LR. The latter experiment tested for the conversion of "food" into carbon dioxide and carbon monoxide, the former for the synthesis of "food" from carbon dioxide and carbon monoxide. The fixation of carbon dioxide into organic compounds is, of course, exactly the process conducted by plants in photosynthesis. Many plant-like microorganisms can perform the step even in an oxygen-free atmosphere. Indeed, as we have seen, it is the way Earth got an oxygen atmosphere.

Why then did the test give negative results with Earth soils? It was not due to a failure in design or concept, for the result was expected. The response failed because of the absence of moisture. The instrument did have the option of allowing water vapor to moisten the test soil. When water vapor was admitted into the test chamber on Earth, the microorganisms, which had been in a quiescent state, responded by fixing carbon dioxide. Soils containing photosynthesizing organisms, such as algae, gave a very strong positive response. If photosynthesizers were scarce or absent, a weak response was still obtained. There are a number of dark processes, common in terrestrial life, by which carbon dioxide is transformed into more complex organic compounds. The energy for these processes is obtained from combustion of foodstuffs, using the chemical energy of the food as the source of the free energy. These dark processes are about one thousand times weaker than photosynthesis, but can still be detected. Barren or sterilized Earth soils gave no response to the test.

Of the three Viking life-detection tests, the carbon assimilation experiment was the one that had been most carefully considered from a theoretical point of view. It was therefore quite fitting that the team leader was Professor Norman Horowitz of the California Institute of Technology. In Chapter 8, we have already considered certain of Professor Horowitz's views concerning the requirements for life. He has published many papers on the properties and constraints of Earthlife. We interviewed him in his spacious and impressive office on the university campus. A huge globe of Mars dominated one corner of the room. Professor Horowitz was serious, gray, dignified, and skeptical.

The carbon assimilation test was the only one that attempted to detect Martian life under conditions approximating those on the planet, Professor Horowitz explained. Experiments that attempted to culture Martian organisms under Earthlike conditions ran a considerable chance of failure, he pointed out. This pitfall was avoided here. Care had been taken to avoid false positives. Professor Horowitz and his coworkers had discovered the reaction, mentioned earlier, in which short wavelengths of ultraviolet light caused the synthesis of organic compounds in a simulated Martian atmosphere. These wavelengths were filtered out from the lamplight used. There was no chemical reaction known before the Viking mission that could have given a false positive result. A false negative was, perhaps, a stronger possibility. While it was reasonable and logical that Martian organisms might use the proposed metabolic pathway, it was by no means necessary that they do so, and other pathways for metabolism could exist. The temperature of the experiment was significantly higher than those normally found on Mars. This was not what Professor Horowitz wished, but it was necessary in order that the other experiments, using liquid water, could function. The other conditions, on the other hand, resembled those normally present on Mars.

When the carbon assimilation experiment was actually

run on Mars, a number of soil samples from the two lander sites were examined. With one exception, all gave a positive response. The responses were weak but definitely significant. They resembled the responses given in the presence of moisture by nonphotosynthetic organisms on Earth. The magnitude of the responses varied considerably from sample to sample. Some runs had been conducted with the lamp off, and others with water vapor added. The total number of positive runs, five, was not enough to allow the variables to be fully separated. It was fairly clear, however, that light and water had no dramatic effect on the response. Preheating the sample for three hours at 170°C destroyed much but not all of the activity. Heat for two hours at 90°C had little effect on it.

Thus, the experimental results afforded an authentic dilemma. A reaction that should have been produced only by living things was not eliminated by heat that would kill those organisms. At least one assumption was wrong. Either an unsuspected chemical reaction on the surface of Mars was causing carbon assimilation, or a type of organism existed there that was more heat-resistant than known Earth organisms. It should be noted that some microbial survival was observed in the LR experiment in terrestrial soils heated at 160°C for two hours. The additional survival capacity required of a Martian organism to account for the carbon assimilation results would be unprecedented but not extraordinary.

The chemical explanation of the results, but not the biological one, could be explored in simulation attempts on Earth. In post-Viking experiments by Professor Horowitz and his colleagues, chemical conditions were found that yielded carbon assimilation, but a duplication of the Viking results was not achieved. There was difficulty in selecting a suitable model for Mars soil. The results of the carbon assimilation experiment remain ambiguous and await further data, which could best be obtained from a new Mars probe.

MARS TODAY: IS THERE LIFE AFTER VIKING?

The Viking instruments represented the first ambitious effort of mankind to detect extraterrestrial life directly. The life-detection apparatus has now shut down, its tale is complete. How will the results be interpreted? A number of divergent opinions have been put forth by interested scientists. The position that we wish to argue is our own. We believe that efforts to draw definite conclusions, positive or negative, from the existing data are intellectually unsound. As it has been for centuries past, the question of the existence of life on Mars remains open. We must further recognize that even a completely negative set of Viking results would only have excluded the presence of carbon-based life, similar to terrestrial life, at the two sites examined. The greater question of whether life exists on Mars would by no means have been settled.

Let us consider the actual results. The organic-analysis experiment found no carbon compounds. Up to a million Earth-type microorganisms per gram of soil could have escaped detection, however. The biology package had three separate experiments, based on different assumptions and apparatus. Each had criteria, established in advance, for positive and negative results. A positive result on any one was to be considered significant, regardless of the others. One experiment gave a positive result. The other two gave unexpected results that satisfied neither the positive nor negative criteria, and do not allow definite conclusions.

On the whole, the biological experiments pointed in a positive direction, and the organic analysis in a negative one. There had been advance expectation that just the reverse might be found: plentiful organic compounds but no sign of metabolism. It was as if we expected to find instruments but no music, and instead encountered music but no instruments. We are left with apparent contradictions and a feeling of confusion. For the authors, and for many scientists we have known, such a state of affairs is a familiar, almost commonplace one. We say, "How interesting. Let's collect more data," and proceed to unravel the snarl. In the interim, although we may have our own secret convictions on the subject, there is no rush to put them into print.

The tradition with Mars, as we have seen, has been otherwise. Individuals have used each new set of results as a pretext to shape the planet to their own psychological needs. In the present case, a majority of those who have commented on the Viking results have issued either carefully qualified or flatly declarative statements that life does not exist on Mars.

For those firmly wedded to the idea that life equals Earthlife, the issue was fairly well settled before the Viking rockets lifted off. If Earthlife could not adapt to the cold, dry valleys of Antarctica (the vegetation in the rocks had not yet been discovered), why expect it on Mars? The organic-analysis data then settled the question beyond doubt. The biological results represented a nuisance to be explained by the best ad hoc chemical interpretation that was at hand. If none was available, we could be sure that one would be found sooner or later.

For an example of the pessimistic point of view, we may quote an article by Professor Horowitz: ". . . at least those areas on Mars examined by the two spacecraft are not habitats of life"; though elsewhere in the same article he cautions, "Until the mystery of the results from the pyrolytic release experiment is solved, a biological explanation will continue to be a remote possibility." An even more absolute, though quite poetic, viewpoint can be found in an address by Dr. Lewis Thomas, as cited in *The New York Times*:

Mars from the look we've had at it thus far is a horrifying place. It is, by all appearance, stone dead. It is surely the deadest place any of us have ever seen, and it is hard to look without wincing. Come to think of it, it is prob-

ably the only really dead place of any size we've ever caught a glimpse of, and the near view is incredibly sad.

This is certainly an exaggeration, since Moon, which men have really seen, appears more lifeless than Mars, whatever the results from Viking. But, as we have observed, there are other reasons to quarrel with Dr. Thomas's statement. The results from Viking are ambiguous.

Some writers have chosen to attack the inconsistencies in the chemical explanations of the Viking results. The biological alternative is then victorious—by default. The conclusion drawn by Robert Jastrow, a well-known physicist and writer of popular science, was that "although the Viking experiments have contradictory elements, they seem to indicate that life, or some process closely imitating life, exists on Mars today."

The problem is that neither the chemical nor the biological point of view holds up well at the present time. How does one choose between alternative poor explanations? To do this the antilife forces have invoked Ockham's razor. This tool of logic, named for an English bishop, has found wide application in philosophical debates. It states that the simpler of two competing explanations should be chosen when both serve to explain a particular set of data. The good bishop failed to leave instructions to deal with the situation in which both hypotheses are unsatisfactory. In the chemical explanation, we must assume that the soil of Mars contains a diverse set of oxidizing and catalytic agents, none of which has been confirmed by any Viking experiment not in the life-detection package. Some of them release oxygen on contact with water, others perish conveniently at 50°C. Their net effect is to mimic, in all cases, a biological response. Compounded with them we have an unspecified carbon-fixing reaction which had completely evaded prior attempts at detection.

In order to ascribe the results totally to biology, we must

invoke organisms that flourish under conditions of cold and aridity which arrest the metabolism of Earthlife. These organisms maintain themselves and their debris in amounts sparse enough to evade detection by the GC-MS. They either do not respond to the conditions of two of the three biology tests, or, alternatively, they do so and consist of a very mixed population of organisms, some perishing at 50°C, others resisting heat at 160°C.

You, the reader, may decide to choose between these hypotheses. We decline to do so on the grounds of insufficient data. Sometimes the wisest option is not to have an opinion until more facts are available.

IF THERE IS LIFE ON MARS, WHAT IS IT LIKE?

The question of life on Mars is still open. This does not mean that we have learned nothing in a hundred years. A much more constrained but better-defined set of possibilities remains. The dying civilizations, the widespread fields of vegetation, the horrid monsters are all gone as possibilities. All macroscopic beings are probably gone. Quite recently, though before the Viking experiments, Carl Sagan observed, "There is no reason to exclude from Mars organisms ranging in size from ants to polar bears." If they exist at Chryse or Utopia, they are quite modest, for they have scrupulously avoided detection by the television cameras on the landers during their operation. Of course, we may not have looked carefully enough. Gilbert Levin feels that the color pictures released did not compensate adequately for the characteristics of the film. When balanced, definite green areas could be seen on the rocks that resembled patches of lichen on Earth rocks. "It could be that the Emperor has clothes," he said proudly.

Perhaps a surprise of this type is waiting for us. But at any event, no obvious signs of life were observed by the Viking cameras. The search for life on Mars has been reduced to a search for microorganisms. This is not surprising. In our current view of the development of life on Earth, billions of years of active evolution were necessary before complex organisms developed. A search for life on Earth through most of its history would have discovered only microorganisms. The planet Mars, at the present, is a cold, arid place, with a thin atmosphere and little or no liquid on the surface. It may have been so for billions of years. The absence of water, in particular, makes these conditions poorly suited for the early development of life of the kind we know.

There are tantalizing hints, however, of a distant Martian past in which conditions were much more similar to Earth's past, with a thicker atmosphere and substantial amounts of liquid water on the surface. Under those conditions, organisms similar to primitive Earthlife might have evolved and flourished. As conditions worsened, subsequent evolution would have been devoted to adaptation and survival. The time and space needed for trial-and-error evolution to more complex forms may never have been available.

Several scenarios remain for the existence of microorganisms on Mars, according to the interpretation that is placed on the Viking results:

(1) Active life was detected in the samples collected by the Viking lander. This life is widespread on Mars.

If so, these life forms are adapted to the actual surface conditions on Mars: cold, radiation, little water, and an active, probably oxidizing soil. Such organisms would need to concentrate water from available sources. Some water would be available as frost, other water would be bound to minerals in the rocks. Carl Sagan and Joshua Lederberg have termed such organisms petrophages and crystophages (Greek for rock eaters and frost eaters), respectively. Energy would be needed to extract this water and heat it sufficiently to keep it liquid within the organism. The efforts to explain the Viking results in terms of chemistry have ascribed a number of unique chemical activities to the Martian soil. If a chemical explanation is accepted for a *portion* of the results, the same activities, ironically, could supply an energy source for living organisms. It would be quite reasonable, for example, to ascribe the LR results to life and the oxygen release to chemistry. Organisms could then survive by scavenging organic compounds made by photochemical processes (and by the decomposition of past life) and combining them with soil-derived oxidizing agents to furnish energy. A number of energy schemes can be imagined, depending on the actual characteristics of the surface (which are unfortunately unknown). The organisms would need to protect themselves from the damaging effects of solar radiation, and to retain heat and water. A thick protective cell wall, or coat, could serve all these functions. The same coat might make the release of organic compounds on heating difficult, hindering their detection by the GC-MS. A plausible, if speculative, life-style for organisms can be constructed, within the framework of the Viking results.

(2) Life is widespread near the surface of Mars, but is not present in the actual Viking samples.

The lichen postulated by Gilbert Levin to be on the rocks near the landers would be an example of this type of life. Unfortunately, no samples were taken from these patches, so the results of the Viking analyses do not apply to them.

In the cold, dry valleys of Antarctica, where stretches of soil lie barren of life, vegetation yet survives within cracks in the rocks. Similarly, microorganisms on Mars might choose to live within the rocks or deep in the soil, and so protect themselves from the harsh radiation and climate. A number of metabolic schemes by which microorganisms deep in the Martian soil could derive energy from the Sun via chemical reactions have been summarized by Benton C. Clark.

(3) Life exists actively in specialized microenvironments on Mars.

The Martian surface is quite heterogeneous, with its volcanic peaks, ice caps, dry river beds, and jumbled chaotic terrain. Within this variety, there may be oases suitable for the survival of organisms not too different from Earthlife in their needs. For example, underground sources of heat might melt the permafrost, and the water produced might be trapped in fissures or caves. We have mentioned that high concentrations of water vapor in some localities have led certain scientists to suggest that water may exist temporarily on the Martian surface. Organisms could survive in such niches, and become dormant in spore-like forms when the niches disappeared. The development of such organisms, especially during the presumed earlier periods on Mars when the climate was more clement, does not seem implausible. The spores of some Earth bacteria might be able to survive in the present conditions on Mars. These bacteria have not even faced the conditions that would have put pressure on microorganisms to evolve in this direction. Indeed, it seems inherent in the logic of Darwinian evolution that if there were microorganisms on Mars in earlier times, they would have developed spores of this type. These spores would be dispersed by winds over the surface and be activated when they encountered suitable conditions. If this were the case, activation of spores may have occurred in the Viking LR and GEX experiments and might have accounted for the results. The absence of active life at the Viking sites would have precluded the debris needed for detection by the GC-MS.

The above accounts are, of course, speculative, as are all detailed explanations of the Viking biological results. To proceed further, we will need fresh data.

The direct exploration of Mars since 1961 has brought us much new information about the planet, although we have not proceeded in a straight line in our understanding. We have not yet obtained enough information to answer most of the questions we have about Martian life, including the essential ones:

1. Has life existed in the past on Mars?

2. Does life exist now on Mars?
3. If so, where does it exist and what kind is it?

A decade ago the detection of extraterrestrial life was considered to be the most important objective of the unmanned space program, with Mars as one of the primary targets for exploration. In our opinion, little that has occurred during the intervening years has altered that value judgment. It would have been superb if the one-shot Viking mission had settled the issue of life on Mars. For this reason a number of writers are eager to assume that the issue has been settled. This is simply not the case. Even absolutely negative results at the two sites could not have excluded life on the planet, and the actual results were by no means negative. It would be an injustice to those who have worked so hard to provide us with the information we now have if we were to stop at this point, substituting unjustified dogmatism about the meaning of this data for an effort to gather essential new information. We believe that the long and convoluted history of human interest in life on Mars provides ample reason for continuing our program of exploration until these questions are settled.

HOW DO WE CONTINUE?

Two lines of thought have emerged in planning for the next mission. It is obvious that the ambiguities of the Viking results could be resolved if the samples were available for investigation on Earth. A variety of chemical and biological techniques would be used, rather than the limited, constrained manipulations possible at the lander sites. Such varied techniques were applied extensively to lunar rocks and left very little question about their nature. A limited sample return mission to Mars would be one of the least expensive options for the continued exploration of the planet.

But what if the wrong sample were collected? One might

be taken whose properties differed from those reported by Viking. Even if the same type of sample were acquired and the Viking results proved to have chemical causes, it would still be possible that life existed in other types of environment on Mars.

Missions that return samples would also raise the difficult question of sterilization. If the possibility of life on Mars exists, then there also exists a remote hazard of Earth being contaminated by returned Martian organisms. Sterilization of the samples before return would diminish their scientific value, while extensive quarantine precautions would increase the cost of the mission.

One alternative to immediate sample return is to continue the remote exploration of the Martian surface. This option would expand our knowledge on the environments of Mars and prepare the way for the intelligent selection of a sample for return. This procedure has been recommended by a Mars Science Working Group appointed by NASA and by the Committee on Planetary Biology and Chemical Evolution of the National Academy of Sciences. The proposal of most interest for the detection of life recommends the exploration of the surface by a mobile rover (Plate 15). The landing site would be selected for safety, and the rover would then travel about in search of interesting local environments. If this option is chosen, we urge that the rover collect a variety of data, rather than perform very specific tests based on the assumption that life on Mars is like life on Earth. It would be very useful if an improved, more sensitive GC-MS were included. This instrument could be readily adapted to carry out simple metabolic tests, such as the GEX, under Martian, rather than terrestrial, conditions. An automated laboratory capable of performing and analyzing chemical reactions would enable us to understand more about the oxygen release and other chemical phenomena reported by Viking. It is possible that the rover would locate soils richer in organic compounds than the ones sampled by Viking. If so, and if the simple metabolic

Plate 15.   An unmanned rover proposed for future explorations of the surface of Mars. Such a rover would allow for the investigation of many different environments in the same mission.

tests were encouraging, these soils would be prime candidates for sample return. Our curiosity might even be provoked enough, and our ambition sufficiently stimulated, to consider a manned landing on Mars. This enterprise would provide a fitting culmination for our long involvement with our neighboring planet.

Mars, of course, is just one of several worlds in our solar system where life may exist. Others that have been less explored are even more promising as homes for life. We will consider them in the next chapter.

Fable 3

The results of the fifth space probe of minor planetoid three were being described at the Jovian Conference on Space Research. Sarpedon, the chief scientist in charge of the probe, reported on it:

> *The probe passed through the thin atmosphere of planetoid three successfully. From the experience that we gained by previous unsuccessful probes, we were able to construct this probe out of special materials that could resist the extreme environment at the gas-liquid interface of the planetoid. The highly oxidized outer coating of the probe enabled it to avoid the fate of probes number one through four, which rapidly combined with a toxic gas in the planetoid's atmosphere. When the probe reached the interface, it was subjected to the chemical action of the hydrogen-oxygen liquid compound that forms the main component of the inter-face. This gradually removed the oxidized protective coating of the probe and so exposed the inner machinery to the toxic atmosphere. As a result, only seventy-two minutes of data were obtained. But this data is enough to confirm the previous opinion of the best scientists—that life is impossible on such planetoids.*

> *If the toxic atmosphere and liquid surface were not enough to show this, an immense flux of deadly radiation of optical light was detected at the surface, which was hardly screened by the thin atmosphere. This radiation can dissociate many chemical compounds that*

are essential to life, and is more intense at the interface of planetoid three even than in outer space near our planet. Also, the temperature at the interface is as low as that in the uppermost levels of our planet. This means that chemical reactions proceed very slowly, and life processes would be extremely sluggish, if indeed there has been time enough for life to evolve there. Finally, none of the complex molecules with which we associate life could be detected at the interface. A sample of the liquid region showed the overwhelming part of its composition to be oxide of hydrogen, with small amounts of dissolved sodium chloride and other metallic salts. There are minor traces of dissolved oxide of carbon, as well as traces of volatile carbon compounds of a type not known on Jupiter. One mobile subprobe was lost in an unknown way, apparently falling into a floating mixture of hydrogen oxide with solidified and nitrogenized carbon compounds. The high temperature of the probe eventually melted this mixture, but not until the probe had been dissolved and oxidized. Small amounts of solid material from the interface were re-covered by another subprobe and placed in a nutrient solution containing essentials of life such as hydrogen cyanide, at an absolute temperature twice the normal value at the interface. At first, the solid material reacted chemically with the nutrients, liberating various gases. But after a short time, the reactions stopped and no further activity was observed. The unwillingness of any hypothetical organism to use rich nutrients is a serious blow to the belief that planetoid three is a home of life.

On the basis of these results, it appears safe to con-clude that planetoid three is not a place where life can exist, and no further biological probes of that planetoid are warranted. Our future studies of the minor planetoids should concentrate on planetoid two, whose thick atmosphere and high temperature at the interface make conditions there much more similar to those on our own world, the only one that we know is hospitable to life. Perhaps life, as we know it, can exist (if only in an attenuated form) on the second planetoid from the central star, but surely not in the wholly alien conditions of the third planetoid.

Sarpedon stopped burping spurts of hydrogen sulfide, which was his method of communicating with his fellow scientists. They, in turn, signaled their approval of his conclusions by producing small pulses of heat, intense enough to boil some of the magnesium chloride crystals contained in parts of their bodies. The result was a small train of bubbles in the dense hydrogen surrounding them all, forming a beautiful but transient pattern pleasing to the speaker and audience alike.

On planetoid three, known to a few of its inhabitants as Earth, countless living things were being born, existing, and dying every second, unaware of the negative verdict about their possible existence which had been rendered by Jupiter's leading scientists.

Chapter 10

Life on Small, Cool Objects

Almost all the matter in the Universe exists in the stars or as isolated atoms and molecules in the depths of space. If the generation of life is an important process in the Universe, then it is in those habitats that the great mass of living things will be found. The smaller bodies that we call planets, the even smaller satellites circling them, and the related objects (such as comets and asteroids) which orbit the stars make up only a tiny fraction of the matter in the Universe. Yet we have a special reason to be fond of these objects; we live on one.

The one on which we dwell—Earth—is the only such object on which we have had a chance to perform truly extensive observations. It is a worthy subject for study, as it contains an atmosphere, large oceans, and a solid surface. We tend to take these features for granted, yet many planets lack one or more of them. As we will see, some planets are just a chunk of rock, lacking seas or atmosphere. Others may be mostly or entirely made of gas or liquid, with little or no rocky core.

Yet, as we have seen, our knowledge even of Earth and the life forms that may exist on it is incomplete. We have not taken full inventory of the creatures in our own biosphere. Other areas, such as the interior of the Earth, have been inaccessible to direct examination. We know even less details about other worlds in our own solar system. Our stock of knowledge has been increasing in the past decade, partly through such spectacular feats as flybys and spacecraft landings, and partly through new measurements taken from Earth of the radiation that other worlds emit. We have learned that the various bodies differ greatly in their conditions, and therefore in the kinds of life that may exist in each locality. A type of life that is appropriate to one environment, such as Earthlife, could not develop or easily survive in another environment. Yet many places in the solar system have their own possibilities for the evolutions of complexity, and there are reasons to expect that the planet Earth is not a unique site for life in our own solar system.

There are, of course, nine known planets in the solar system, varying in size and composition (Table 10-1). The solar system as a whole is a little less than five billion years old, not much older than Earthlife. It is thought to have condensed out of a dense cloud of the type observed now in other parts of the galaxy. The process by which the individual planets were formed is still not well understood. The difference in the planets as they now exist is thought to be a consequence of varying conditions of temperature and gravity at the different locations where the protoplanets formed.

A few of the planetary satellites are comparable in size and in their possibilities for life to the planets themselves. The properties of several large satellites are described in Table 10-2. Also found in profusion in the solar system are the comets, low-mass collections of dust and solids that move in very elongated orbits around the Sun. Other low-mass bodies are the asteroids, an array of irregular rocky bodies that exist mostly in the region between Mars and Jupiter. Some astronomers believe them to be the remnants of a planet that never formed.

So far as we know, the planets and their major satellites have existed in their present form for several billion years.

Table 10-1
PLANETS OF THE SOLAR SYSTEM

Planet	Mass of Planet Divided by Mass of Jupiter	Diameter (meters)	Average Distance from Sun (meters)	Surface Temperature
Jupiter	1	14×10^7	0.8×10^{12}	$-120°C$ $(153°K)$
Saturn	1/3	12×10^7	1.4×10^{12}	$-160°C$ $(113°K)$
Neptune	1/20	5×10^7	4.5×10^{12}	$-220°C$ $(53°K)$
Uranus	1/25	5×10^7	2.9×10^{12}	$-200°C$ $(73°K)$
Earth	1/320	1.2×10^7	0.15×10^{12}	$-80°C$ coldest $+50°C$ hottest
Venus	1/400	1.2×10^7	0.1×10^{12}	$+470°C$
Mars	1/3000	0.7×10^7	0.2×10^{12}	$-140°C$ coldest $+20°C$ hottest
Mercury	1/6000	0.5×10^7	0.06×10^{12}	$-170°C$ coldest $+400°C$ hottest
Pluto	1/130000	$.27 \times 10^7$	6.0×10^{12}	$-230°C$ $(40°K)$?

The planets of our solar system, listed in order of decreasing mass. The first four planets have no solid surfaces, so the temperature has been given for an atmospheric layer at a pressure corresponding to Earth's at sea level. When a sizable temperature range exists, as on Earth, maximum and minimum values are given. The question mark indicates that the value listed is uncertain.

Table 10-2

SOME LARGE SATELLITES

Satellite (Planet It Orbits)	Diameter of Satellite Divided by Diameter of Earth	Surface Temperature
Titan (Saturn)	.45?	−70°C
Triton (Neptune)	.45?	−220°C (50°K) ?
Ganymede (Jupiter)	.42	−140°C
Callisto (Jupiter)	.38	−140°C
Io (Jupiter)	.30	−140°C
Moon (Earth)	.27	−180°C coldest +110°C hottest
Europa (Jupiter)	.25	−140°C

The largest satellites in the solar system, arranged in descending order of size, and some of their properties. Question marks indicate that the values are uncertain.

There is evidence from cratering on Moon, Mars, and Mercury of a period of frequent collisions between planets and debris left over from the time of planetary formation. These collisions may mostly have occurred in the early days of the solar system, and have diminished in the last few billion years. There are slow physical, chemical, and perhaps biological changes taking place on and in the bodies of the solar system, but the time scale for these changes is in millions of years, so that planetary environments are for the most part fairly stable. This means that many processes can go on in a uniform way without sudden radical changes in conditions. This is probably important for most forms of life, as complex systems tend to take a while to develop and adapt to new conditions. If conditions change too rapidly, the systems cannot adapt at all and perish. In some cases the outer regions of a planet are subjected to environmental effects, such as large amounts of radiation, which would be inimical to Earthlife. However, these effects may be like gentle rain to the creatures that have evolved in that environment. It is not a specific environment or a lack of it that life needs, but a period of environmental stability long enough for evolution to occur, and this seems to have been available in most of the solar system.

Also, deep under the atmosphere or surface of the planets, any life would be shielded by the matter covering it from radiation hitting the planet from outside, just as fish in the deep oceans of Earth are shielded from variations in the temperature of the surface. This is another reason to consider seriously the possibilities for life deep inside planets.

Of the worlds in our solar system other than Earth, we have learned the most about Moon because of its proximity and the manned landings on it. Venus and Mars have been the targets of successful unmanned landings, there have been flybys of Mercury and, successively, of Jupiter and Saturn. One of the latter probes may go near Uranus some years from now. Our knowledge of the other worlds is based on remote observation. Much of what we say about all of them, especially about their insides, is conjecture. It is informed conjecture, based on theoretical science, but still subject to error. Almost certainly, the present scientific picture of the planets has serious mistakes in it, which will be corrected by additional observations and calculations.

Despite these uncertainties, we will use existing knowledge to speculate further and suggest where life may be found in the solar system. We hope that our suggestions will stimulate scientists and the public to investigate the solar system and find out more about what may exist there.

THE WORLDS OF OUR SOLAR SYSTEM

In exploring our solar system in greater detail, we will again find it convenient to use COSMEL. When we studied the architecture of life, we used its enlarging mode to the +2 level and then reversed directions. Let us now return to our starting point in New York and push the +4 button. We step out to see a view similar to the one visible from an airplane window on transcontinental flights. We stand over 15,000 meters high with our heads above the clouds. There is a blue haze on either side of us, but the sky darkens above us. Below us, through openings in the clouds, we see the grids of the city streets and the winding paths of rivers. Let us pause and use this vantage point to examine our planet in greater detail. This will not be difficult, as with our present height, we could walk around the Earth in less than an hour at a brisk pace.

On this +4 level we find that the highest mountains are about one-half our height, while the deepest ocean basin is a pool about the same in depth. The average ocean is less deep, perhaps reaching up to our ankles. Three-quarters of the surface is covered with water, and only a quarter is solid land. The land surface is not homogeneous. About 5 percent of it, almost all near the South Pole, is covered with water ice. The uncovered land contains some areas of sand, some naked rocks, and some regions consisting of permafrost, a semifrozen mixture of rock and ice. But much of the land surface is covered by the thin, thin layer of organic chemical compounds mixed with liquid water that we know as Earthlife. On level +4 the tallest example of Earthlife is but one centimeter high. If Earthlife were spread out uniformly over the surface of the planet, it would form a layer no thicker than a sheet of paper. Earthlife is truly a two-dimensional phenomenon. Yet viewed from our present perspective, about what we would see from a high-flying jet plane, this thin layer of life is one of the more prominent visible features of Earth. If we look from a height that is one hundred times greater, say from the +6 level, the viewpoint of an orbiting satellite, almost all trace of life disappears unless sensitive instruments are used.

The land surface material is in the form of solidified mixtures of iron, aluminum, and silicon oxides, and smaller amounts of compounds of other elements such as carbon.

The oceans are over 95 percent water, with a few percent of compounds of other elements, principally sodium chloride, dissolved in it. All these surface materials have a lower average density than the whole Earth, as the heavier elements tended to sink to the center soon after the Earth formed, when it was still molten.

When we stand on Earth on level +4, the atmosphere is mostly below our head, although traces of it extend far out into space. This atmosphere contains about 80 percent nitrogen molecules, 20 percent oxygen molecules, and small amounts of other substances. This was not the primordial atmosphere of Earth, which contained no free oxygen and substantial amounts of carbon dioxide, as Mars' atmosphere does today. The present atmosphere has been produced over billions of years, largely by the action of Earthlife. It also contains about 1 percent water vapor, although this amount varies a great deal from time to time and place to place. This water vapor, when it condenses as dew or falls as rain, is an essential source of water for land Earthlife, which otherwise would need to find an ocean every time it needed a drink. The density of the atmosphere at the surface is about 10^{-3} that of the surface material itself. So, the surface is a real boundary between two very different regions, a situation quite different from that on some other bodies of the solar system.

As we have noted, the main energy flow to Earth is sunlight. But some energy also reaches the surface from the other direction in the form of heat flowing from the deep interior, where it is produced by decaying radioactive elements. This heat flow amounts, on the average, to about 10^{-5} as much energy as sunlight. However, in several places, such as hot springs, heat flows can be more significant and can be the main energy supply for some forms of life. Smaller flows of energy include lightning discharges and the high-energy particles called cosmic rays.

The various energy flows interact with the chemical compounds in the atmosphere, oceans, and surfaces to produce a rich variety of chemical reactions. The most interesting, as far as Earthlife is concerned, are those involving the compounds of carbon, nitrogen, hydrogen, and oxygen. Since we have already discussed some of these reactions and their roles in Earthlife in detail, let us take instead a brief look at Earth on a larger scale.

We will return to COSMEL, adjust its screen and distance gauge so that they point about ten thousand kilometers out into space, and push the +7 button.

We are now larger than the Earth itself, which hangs before us, a beautiful four-foot, blue, cloud-covered globe. We appear to stand on empty space, though we can imagine that our model contains a transparent floor positioned just below the orbit of the Earth around the Sun. A good analogue for the Earth's structure is a giant spherical peach, slightly wet from just being washed. The atmosphere and surface are insignificant parts of the whole Earth. A layer less than one millimeter thick on the +7 level represents all the water in the oceans, while a slightly thicker layer above the surface contains most of the atmosphere. The differences in elevation between the highest mountain and the ocean depths are no more apparent on level +7 than is the fuzz on a peach.

Extending about one-half centimeter beneath the surface on level +7 is a thin layer composed mostly of solid rock with occasional pockets of molten rock. This is the crust. Under the crust, like the pulp of a peach, is a region of dense, still mostly solid rock known as the mantle, extending about one-quarter meter to the center. The center of the Earth is a still denser region, composed of iron and other metals, partly liquid and partly solid—the peach pit. This core is at a temperature of 6000°C, hotter than the surface of the Sun. Only the intense pressure of the whole mantle above it keeps the material from vaporizing. The liquid part of the core is not stationary, but flows in immense currents generated by the rotating Earth. It is believed that these currents in the core produce Earth's weak magnetic field.

Now that we have seen something of the outside and inside of the Earth, and while we are still on level +7, let us turn and look outward from our position near Earth. The Sun itself and the full Moon shine against the black backdrop of space. We will approach the closer body and walk out across empty space toward Moon. We approach it quite rapidly, as it is less than a city block away at our present level. It is a large pock-marked ball, about thirty centimeters in diameter.

We have come here more to pay homage to the first site chosen in the manned exploration of space than to search for life. Thanks to the imagination and dedication of American Presidents and Congresses and half a million workers, twelve men have walked Moon's surface, studied conditions there, and brought back rock samples to Earth. The results gave no sign of lunar life. No indication of such life was found either on Moon itself or in the samples that were returned and analyzed.

The surface of Moon is an unlikely site for life, as unfavorable as any in the solar system. A plentiful supply of energy impinges on its surface, but falls upon no medium, no gas or liquid, in which a suitable interaction of chemicals could take place. Compared with Earth, Moon is significantly depleted in carbon and hydrogen and almost devoid of water. Moon is smaller and has less surface gravity than Earth. Because of this, any light elements and compounds Moon may have possessed would have escaped easily.

Although we have no shred of evidence for lunar life, we must recognize that only the surface and the upper few meters of the crust have been studied. Conditions inside could be quite different. During the lunar day (two weeks on Earth), there is a flow of heat from the surface to the interior. This flow is reversed during the lunar night. The flow is sometimes 10 percent as intense as that of sunlight on the surface, and could be a source of free energy for a subsurface life. The possibility of such life would exist if volatile materials, such as water, have persisted at a few meters depth. Even if the odds for such life are quite slim, it would be worthwhile to look for it by boring deep holes into the lunar crust whenever human beings return to Moon to stay. But right now we will move on. We need not struggle with such slim possibilities when much more likely opportunities lie ahead of us. Moon lacks an atmosphere, carbon compounds, and water. So let us set our sights on a world where all these items, and more, are available in abundance. We head for Jupiter.

Although we reached Moon in minutes from Earth at the +7 level, it would take most of a day to hike the sixty-three kilometers to Jupiter on this level. To save time, we push the +9 button on COSMEL and survey the scene again. Earth has become a disc the size of a shirt button, Moon as big as a pinhead, and they are half an arm's length apart. The Sun, a glowing yellow globe not quite our size, is about two city blocks away. We turn our backs on it and start to walk out toward Jupiter, four times as far away.

The trip is not without interest. We pass Mars en route (we have thoughtfully placed most of the planets on the same side of the Sun in our model). Mars is a tiny copper-red disc, half the size of the shirt button Earth, at our present level. There is no need to make a stop here, using COSMEL. Mars has already been visited in reality by two unmanned landers of the Viking project, as described earlier.

Many scientists have concluded that Mars is the planet with conditions closest to those on Earth (which is debatable) and then leaped to the Earth-centered conclusion that it is therefore the sole plausible site for extraterrestrial life in the solar system. We disagree, and will continue on our trip to point out more favorable locations for life.

JUPITER: WHERE THE ACTION IS

Jupiter is a significant body even if we approach it on the +9 level, where we are larger than the Sun; at that level the planet is the size of a cantaloupe. To better appreciate its magnitude relative to other planets, we will return to the +7 level, where we are the size of Earth. Jupiter is now a mammoth ball about fifteen meters high, taller than a three-story building.

Jupiter in fact contains more than two-thirds of the mass in the solar system, apart from that in the Sun, and contains most of the matter in the solar system that is in the form of atoms or molecules. It is an immense planet compared with Earth, with over one hundred times as much surface area and three hundred times as much mass. The ratio of areas is about that between the United States and Austria. Its extreme size suggests that most of the interesting action on the planets in the solar system may be going on there. This suggestion is reinforced by Jupiter's appearance. Its surface is a turbulent mass of clouds arranged in a series of rapidly moving colored bands (Plate 16). Seen from a distance, the pattern of color looks something like the lights of a cosmic discotheque. Set among these colored lights is a unique feature, a large red spot, which has persisted for hundreds of years.

Jupiter differs in more than size from Earth, Moon, and Mars. These, and other inner planets, are not only smaller but denser than Jupiter. They have lost most of the volatile substances such as hydrogen and helium that they originally contained, and so are composed mainly of iron, silicon, and oxygen. The giant outer planets, Jupiter, Saturn, Uranus, and Neptune, have retained more of their volatile substances and are composed largely of hydrogen and helium.

Our knowledge of the structure of Jupiter is very incomplete and is based on recent Pioneer and Voyager flybys and theoretical calculations. We will explore our model by making a trip to the center of it. At intervals we will use COSMEL at a level useful for inspecting molecules (level −8) to check the nature of the substances present.

In the very outer regions, it is quite cold, perhaps −125°C (150°K). The hydrogen molecules and helium atoms are

Plate 16. Jupiter and its satellite, Europa, as photographed by the Voyager 1 space probe. The sizes are slightly smaller than they would appear on the +9 level of COSMEL.

spaced quite far apart. (The pressure is much lower than that at the Earth's surface.) As we descend, the pressure and the temperature keep increasing. When we have descended perhaps one-tenth of 1 percent toward the center of Jupiter, we encounter clouds of frozen ammonia. It grows darker when we pass below them. (The illumination at the top of the atmosphere was not very bright to start with. The sunlight reaching it is only 4 percent as intense as on Earth.) Fortunately, we have brought our own illumination and can proceed.

We continue to pass through cloud layers, first consisting of salts derived from ammonia, then of ice crystals. We are now perhaps 1 percent of the way in, the pressure is approaching one Earth atmosphere, and the temperature is pushing toward $-100°C$. At about one seventieth of the distance to the center, we reach an area containing water droplets, and the temperatures and densities are comparable to those at the Earth's surface. Hydrogen and helium predominate, but organic molecules and ammonia are also present. These conditions pertain over a band about one hundred kilometers deep (in actual distance), which contains more matter than is present in all Earth's oceans. This zone contains the type of chemical mixture which was once thought to predominate on the early Earth. Many simulation experiments have shown that a complex mixture of organic molecules is formed in such an environment when an energy flow is supplied.

The temperature and pressure in Jupiter continue to increase as we descend below the Earthlike level. Little is known about the exact chemistry in this area, but it is likely that we will pass through various layers, each with its own characteristic temperature, density pressure, and composition. Hydrogen and helium will of course predominate, but other substances will exist to the extent of one part per thousand or so, in the same way that water vapor is a minor addition to the nitrogen and oxygen of Earth's atmosphere. These trace chemicals may, however, segregate themselves into solid or liquid clouds, where they are the main constituent, as happens in our atmosphere.

About 20 percent of the way to the center of Jupiter, the atmosphere is as dense as water and behaves more like a liquid than a gas. Beyond this, the pressure and density continue to increase gradually until a point is reached at which hydrogen goes into a metallic state in which the electrons separate from the nuclei and roam freely. At this stage we have traveled about one-third of the distance to the center of the planet. The pressure may be three million Earth atmospheres, and the temperature near 11,000°C. If we continue in this direction, we eventually reach a small, rocky iron and silicate core in the center of the planet, at a temperature of perhaps 30,000°C. Even this central core may not be solid because the behavior of metals and rock at very high temperature and pressure is not well known (See Fig. 25).

As we retrace our steps to the surface, we may puzzle over the energy source that serves to heat the planet. In fact, it does more than that. Jupiter emits over twice as much energy as it receives from the Sun. It is a type of mini-sun in itself. One possibility for the energy source is the decay of radioactive atoms in the center, though this may furnish a bit less energy than is needed. Another possibility is that Jupiter is contracting very slowly in size, making its gravitational energy smaller. This loss in gravitational energy is converted into heat. A decrease in Jupiter's size by a millimeter each year would be enough to supply the needed energy, and it would be a long time before we noticed the shrinkage.

We have seen that Jupiter offers a vast variety of environments, each of which could harbor a characteristic form of life. Let us imagine what some of these life forms might be like. Some could be relatively familiar to us, others would be utterly strange, as the conditions under which they live would be vastly different from those on the surface of Earth. All life forms would have to face a number of problems. These include having a usable energy supply and a method

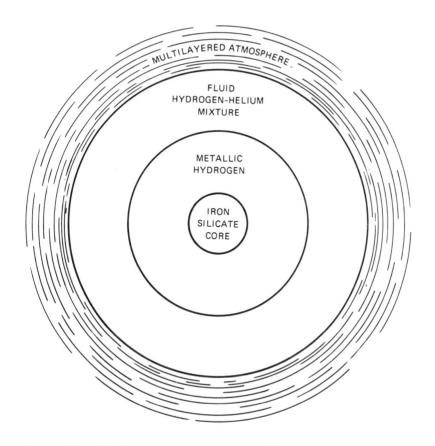

Figure 25. The interior of Jupiter is believed to contain a relatively small central core made of iron and rock, perhaps no bigger than the Earth. This is surrounded by a mixture of hydrogen and helium in various forms. The region just outside the core contains hydrogen in a metallic form, yet unknown on Earth. Farther out is a fluid mixture of hydrogen and helium. The outermost layers are complex mixtures of gases and liquids, in which a variety of chemical compounds occur at different levels.

of supporting themselves on a planet without a solid surface in order to avoid being transported to other regions whose conditions would be destructive to them. The free-energy sources could include sunlight, the temperature difference existing between different layers of the atmosphere, and flows of heat or radiation from the deep interior.

The main source of free energy in Jupiter is likely to be sunlight, at least in the upper atmosphere. The sunlight probably does not penetrate much below the level of Earth-like temperature and densities because it is absorbed by the material in its path. However, there can still be an indirect effect, at lower levels, of the free energy in the sunlight. One thing the sunlight does in the upper atmosphere is produce complex molecules. This happens because sunlight, especially its ultraviolet part, can break up the simple molecules present there, producing ions and other highly active particles, which then combine to form complex, energy-rich molecules, in much the same process discussed in Chapter 5 for the pre-biotic Earth. On Jupiter these complex molecules, being denser than the surrounding upper atmosphere, will drift downward into the inner atmosphere. Eventually, they will be broken apart at the high temperatures inside Jupiter. However, on the way down some of them can react with whatever exists between the level at which they are produced and the one where they break up. This process provides a source of free energy. Conceivably, life forms might exist which use such molecules as their food source, similar to some deep-sea fish which feed on plankton that fall from the ocean's surface.

The difference in temperature between the varying levels of Jupiter's atmosphere could be a source of free energy if material could be transported in a cycle from one layer to another. An object could be warmed by contact with the hotter inner levels, and then travel to the colder outer levels where heat would be lost. The difference between the heat gained and lost would then be available as free energy. We have given the name "thermophage" to a life form which repeatedly carries out this process, "feeding" on the heat it absorbs in the lower atmosphere, and "ex-creting" it in the upper atmosphere.

In order for this process to work with a useful efficiency, there must be a substantial temperature difference (of 50°C or more) between the hot and cold levels. This corresponds to a distance of twenty-five kilometers or more on Jupiter, and so the thermophages must range over at least this distance for each meal. It would seem easily possible in the low-density regions of Jupiter's outer atmosphere. A more important constraint may be the amount of free energy that can be obtained from a temperature difference. For objects at Earthlife temperature, the thermal energy content in a gram of material is much less than the chemical energy content of the foods which are the source of free energy for Earth animals. This implies that Jovian thermophages would have to work much harder than animals on Earth do to keep themselves alive, or else be satisfied with a much lower metabolic rate.

A major influence on the forms life may take on Jupiter is the lack of a solid surface. This fact, combined with the continued increase in temperature and pressure as the center of the planet is approached, suggests that Jovian life forms will either have to be able to survive over much greater ranges of temperature and density than occur on Earth, or they will have to support themselves at an approximately constant level in the atmosphere. This is the same problem that we will discuss in connection with living things on the Sun and in the atmosphere of Venus. Any unsupported large object, such as a living creature, on Jupiter would drift slowly downward because of gravity until it reached a region at which the density of the atmosphere is equal to its own. At this level it would remain approximately stationary except for winds or other large-scale atmospheric motions, which would also move the object around. If the organism is like Earthlife, its density is about that of water, and the level at which the atmosphere has this density is many thousands of kilometers into the planet, at a temperature of thousands of degrees. An Earthlife form would not be viable there because of the chemical changes in it produced by the high temperatures. Therefore, Earth-type organisms would have to be able to support themselves in Jupiter's upper atmosphere in order to survive.

One way for an organism to support itself would be for it to contain enough low-density hydrogen gas so that its overall density is that of the upper atmosphere, even though its functioning parts are of much higher density. Such "living balloons" would have skins whose thickness is 10^{-3} or less of their size, in which most of the living matter would be concentrated, while their insides would be filled with rarefied gas. It would be possible for such a balloon to rise and fall by varying its volume if it had a flexible skin. Some ideas about how such living balloons might function on Jupiter have been given by Edwin Salpeter and Carl Sagan. Another means of support would be to take advantage of areas on Jupiter where permanent natural updrafts exist, powered by the temperature difference between the levels of the atmosphere. There is reason to think that such areas, with updrafts of ten meters per second, exist within Jupiter. Relatively dense objects could be supported by these updrafts, although their lives would end abruptly if the wind stopped blowing upward for a while.

There is no known Earthlike equivalent of life forms that could disregard the very high temperatures and densities deep in Jupiter's atmosphere, and so either live permanently at these depths or travel between upper and lower levels. At present we can only reason about such creatures on the basis of our general knowledge of chemistry and physics. At a depth of tens of thousands of kilometers into Jupiter's atmosphere, the density of the "gas" becomes greater than that of familiar solids, and such solids could then "float" on the gas. However, at these depths the temperatures are many thousands of degrees, high enough to disrupt the chemical hydrogen bonds that Earthlife uses. There are stronger chemical bonds which would remain intact even at those high temperatures, and living material that involves only such tighter bonds might be viable deep inside Jupiter.

Perhaps only a tough protective coat would contain these strong bonds, with the inside of the creature being made of molecules with weaker bonds. Analogues to this construction on Earth are known as oysters.

Another hazard that such creatures would need to avoid is that of being melted and boiled at the high temperatures inside Jupiter. While at Earth's surface pressures all materials boil at temperatures below 6000°C, this is not the case at the very high pressure inside Jupiter or even deep inside the Earth. At pressures that are thousands of times greater than at the Earth's surface, the melting points of many solids are increased or the substances change to different physical forms; for instance, graphite into diamond. For example, nine different forms of water ice have been found to exist at very high pressures, and some of these may actually exist inside Jupiter or other planets. Therefore, it is possible that solid or partly solid bodies can survive deep inside Jupiter despite the high temperatures, and that such temperatures do not rule out some form of life utilizing these solid bodies. The high temperatures and densities in the deep atmosphere would tend to accelerate the rate at which physical and chemical processes occur, enhancing the prospects of evolving some form of life deep inside of Jupiter.

Such a life form might use heat flow from the core as its source of free energy. The basis for its storage of order might be the many different physical forms that matter can take at high temperature and pressures. Different spatial arrangements of these various "phases" of matter might become orderly under the proper stimulus of external free energy. But we do not know enough, either about conditions inside Jupiter or about the behavior of matter under these conditions, even to speculate reasonably about these possibilities.

If life is a widespread phenomenon in the Universe, it will very likely be found on or in Jupiter, where so many different environments exist and where the essential factors of matter and free energy are present in abundance. The search for Jovian life will not be easy for us to carry out, but it will almost certainly be the right place to look.

GANYMEDE—AN INTERNAL OCEAN 500 KILOMETERS DEEP

Our tour of the solar system will not stop at Jupiter because other attractions await us, some quite close at hand. As we emerge from the giant globe of Jupiter, still at the +7 level of COSMEL, we see a number of illuminated bodies nearby, against the backdrop of space. These objects are the moons of Jupiter. Four of them, first discovered by Galileo in 1610, are much larger than the rest. In order of increasing distance from Jupiter, they are Io, Europa, Ganymede, and

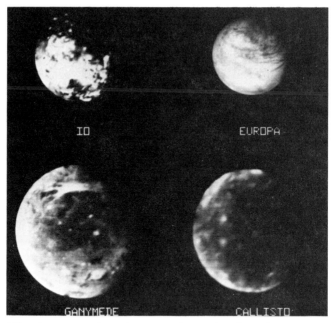

Plate 17. A composite photograph of the four largest satellites of Jupiter, taken by the Voyager 1 space probe. The relative sizes are correctly given in the photograph. The satellites might appear this way on the +8 level of COSMEL.

Callisto (Plate 17). Io, the closest to Jupiter, has an active system of volcanoes, perhaps as a result of the strong gravitational pull of Jupiter on Io's interior. But we do not know enough about Io to speculate about life there. It is interesting to note that one important requirement for life—a suitable fluid medium—is present, as oceans of sulfur probably exist beneath the crust. The other three moons are fairly similar to each other. We do not have time to tour them all, so we will visit the largest one, Ganymede. A brief stroll of about the length of a football field brings us to this moon. Ganymede is an object of planetary size, larger than Mercury. At our level it is about the size of a large beach ball.

Limited information about Ganymede is available from flybys and Earth-based observations. One of its most interesting features is its density, which is quite low compared with that of Earth or Mars. Ganymede must be at least partly composed of much lighter material than the inner planets. It is not large enough to retain hydrogen or helium, and its atmosphere is thin in any event. Water and ice are logical candidates for the light material. There is some evidence that Ganymede (and Callisto and Europa) is made of a combination of water and sandlike material, similar to Earth's crust. Various models for Ganymede exist; the one we present is based on the work of John S. Lewis and his collaborators. We observe that the surface of Ganymede is made of solid ice, at a temperature of about $-170°C$ ($103°K$). The ice is rent by large cracks due to meteor impact. There is little prospect for life similar to that on Earth on the surface, but this does not mean that Ganymede cannot be a haven for life. Beneath the ice crust, which is fifty to one hundred kilometers thick, lies an enormous ocean, five hundred kilometers deep. The situation resembles that of our Arctic ocean, but the Ganymede ocean is vaster. There is twenty-five times as much liquid water under the ice of Ganymede as on all of Earth. Below this ocean is the rocky core, at a temperature that varies from 25°C at the bottom of the ocean to several thousand degrees at the center of Ganymede (See Fig. 26).

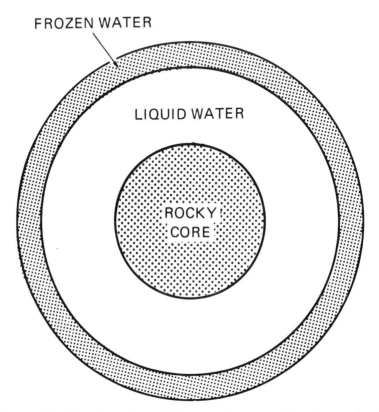

FROZEN WATER

LIQUID WATER

ROCKY CORE

Figure 26. The interior of Ganymede according to a theory described in the text. A small rocky core is surrounded by a deep ocean, primarily composed of water. The surface is a relatively thin coating of solid ice and rock.

Originally, when Ganymede condensed out of the primitive solar nebula, the water was probably chemically combined with sandlike material, as it often is on Earth. When the inside of Ganymede was sufficiently heated by the decay of radioactive material, the interior melted. The various materials separated according to their density, as happened on Earth. The heavy sand fell to the center, forming the core, while the lighter water migrated to the outside. The surface portion in contact with outer space then froze, giving Ganymede its present appearance.

Neither Ganymede's ice surface nor its ocean is pure water. The water contains dissolved impurities of many kinds, just as Earth's oceans do. The precise chemical form of these impurities is unknown, but they may well contain the same elements and simple compounds present in the primitive oceans of our planet. Furthermore, Ganymede's ocean has probably existed in its present form for several billion years. Therefore, this ocean satisfies two of the conditions necessary for life—a suitable material base and enough time for prebiotic and Darwinian evolution to take place.

The crucial factor which may determine whether life exists in the ocean of Ganymede is whether a suitable energy source has existed to drive the matter away from equilibrium. The water is shielded from the feeble sunlight of Ganymede by the ice crust. It is hard to imagine any useful energy getting through to the ocean from above. However, there is another direction from which energy can reach the ocean—underneath from the hot rocky core. The same radioactive decays that originally melted Ganymede are still producing heat in the core, and this heat works its way out to the ocean in various forms. In order to be of use as an energy source for life, the internal heat must reach the ocean in some place in a concentrated form, such as in a volcanic eruption or an upwelling of hot gas. Otherwise, the heat will just raise the overall temperature of the bottom of the ocean slightly, and will not be available as free energy for life. In our present state of knowledge of the internal workings of Ganymede, we cannot be sure whether rich concentrated energy sources will exist under its ocean. Analogies with Earth would suggest that a significant fraction of the energy would emerge in concentrated form at local hot spots, and at those spots, the deviations from equilibrium that are the beginning of life may occur. (The places on the ocean bottom on Earth where hot springs emerge are rich sites for living creatures. These areas derive their primary energy source from minerals in the hot springs, rather than from the Sun.) The exploration of Ganymede's inner sea will be no easy matter. We have not even explored the interior of the Earth at a depth of one hundred kilometers. But the challenge of finding life completely independent of the Sun within the solar system might spur us on to eventually finding a way to look under the Ganymede ice cap.

We will now leave the Jupiter system and move outward. Using the +9 level, we make the .65-kilometer hike in a few minutes and reach the orbit of Saturn. This planet is close at hand, and we return to the +7 level to enjoy the view. Saturn looks somewhat like Jupiter, but is almost 15 percent smaller in diameter and has, of course, a famous set of rings to add to its beauty. Jupiter and Uranus also have rings, but Saturn's are much more spectacular. The internal structure of Saturn and the prospects for life there are probably similar to those of Jupiter. Although we appreciate the view, we have in fact not come to visit Saturn at all, but rather its satellite, Titan.

TITAN—MORE LIKE A PLANET
THAN SOME PLANETS

Titan is one of the two largest satellites in diameter in the solar system (the other is Triton, a satellite of Neptune), and is about 10 percent larger than Ganymede. Its internal structure is probably similar to that of Ganymede, with a thinner ice crust. However, these are not the reasons why we have chosen to visit it. What makes Titan especially interesting in our search for extraterrestrial life is that it has a thick atmosphere, perhaps twenty times as dense near its surface as Earth's atmosphere. This gives Titan an orange color with bluish tints, as observed in the Pioneer 11 flyby. The main constituents of the atmosphere are uncertain. Methane has been detected there, and there may be hydrogen. It is suspected that most of the atmosphere is nitrogen, which cannot be easily detected by its radiation. It is not well understood why Titan should have a thick atmosphere, while a similar body like Ganymede has no significant atmosphere.

The presence of the atmosphere is likely to modify surface conditions on Titan in two ways that are important for possible life. One is that the surface will be much warmer than it would be without the atmosphere. This is due to a process called the "greenhouse effect," which we will discuss when we visit Venus. Very recent measurements of Titan's surface temperature give a value of $-73\,^\circ$C, about as high as that of Mars, which receives fifty times as much sunlight. At this temperature and pressure, there are a number of hydrocarbons that are liquid, and Titan may well have bodies of liquid on its surface.

Another effect of the atmosphere would be the production of substantial amounts of complex molecules by sunlight breaking up the methane and other gases, which then recombine. These large organic molecules eventually fall to Titan's surface, where they either lie or dissolve in whatever material is there. It is possible that there are very large amounts of these organic molecules on the surface, perhaps more than in all the living creatures of Earth.

The main source of free energy in Titan's atmosphere would be sunlight. The intensity of sunlight hitting Titan is only about 1 percent of that reaching Earth. This condition, together with the lower temperatures in the atmosphere and on the surface, suggests a slower rate for chemical and physical processes than on Earth. However, if the atmosphere is much denser than Earth's and there are liquids on the surface, this conclusion need not hold. In a dense atmosphere at Earthlike temperatures, the reaction rate between molecules can be even higher than in water solution. For life in the atmosphere, a possible source of free energy could be the temperature difference between different layers of the atmosphere, as discussed for Jupiter.

The interior of Titan is probably an even more plausible place for life than that of Ganymede because of the known presence of abundant amounts of carbon and nitrogen on this satellite. Several possibilities exist for an internal ocean: ammonia-water, with dissolved organic material; hydrocarbons; or even both in separate layers. As its energy source Titan life could have heat flows from the deep interior, if the life were at the ocean bottom near the core. Alternatively, if bodies of liquid exist on the surface, energy-rich molecules produced on the upper atmosphere could fall and provide an energy source. These possibilities, and our general ignorance of conditions there, make Titan a very attractive candidate for close study. Future space flights to the outer solar system should be planned with this in mind.

We will not venture farther away from the Sun. Uranus and Neptune lie out there, smaller, cooler cousins of Jupiter and Saturn. If life forms exist on these planets, they may be similar to those discussed for Jupiter. But too little is known of Uranus and Neptune for us to model them. Perhaps we will learn something more about them in the twenty-first century. Even farther away from the Sun lies Pluto, about which we know even less. It appears to be more like the rocky inner planets than the gassy outer ones, with a diameter somewhat smaller than Earth's. Pluto is forty times as far from the Sun as Earth is, and its surface temperature may be as low as $-233°C$ ($40°K$). Because of the dearth of information on Pluto, we will not speculate at this time about possibilities for life there. Let us rather turn around and hike back at the $+9$ level, past the orbit of Earth. At about two-thirds of the distance from the Earth to the Sun, we approach the planet Venus and change to level $+7$ to observe it.

VENUS—EVEN HELL FEELS LIKE HOME
TO DEMONS

At first glance, Venus looks quite like the Earth. It is slightly smaller in size and covered with dense clouds. For this reason, some past science fiction writers have speculated that life on Venus might be like past and present life in the warmer parts of Earth. Dense swamps, thick jungles, and reptilian monsters have been described in lush detail. But we know now from modern measurements with radio waves and from several space probes of Venus that the conditions of temperature, pressure, and atmospheric composition make this planet extremely inhospitable to Earth-type life. Apparently the fact that Venus receives about twice as much sunlight as Earth, because it is two thirds our distance from the Sun, has led to quite a different history of the planet and the very different conditions that exist there at present. This is not the only example we will see in which a seemingly small difference in the environment can result in immense differences in the development of an astronomical body.

One significant feature about Venus that has been discovered in recent years is the very high temperature of the planetary surface. A recent Pioneer probe measured a surface temperature of 457°C, or over 840°F, hot enough to melt lead. In fact, such elements as sulfur and phosphorus, if they are present in uncombined form on the surface of Venus, would exist as liquids. Conceivably such liquids could act as a medium for the evolution of chemical life on Venus. It would certainly not be Earthlife, as proteins, nucleic acids, and other key molecules are not stable at that temperature. Other complex molecules, more stable to heat, would be needed. The formation of such molecules depends upon a source of free energy. A low flux of sunlight reaches the surface of Venus because of the thick cloud cover. It might be insufficient for the purposes of life, although even on Earth some algae can photosynthesize with very low intensities of sunlight. Other sources of free energy such as thunderstorms, strong wind, and volcanic action might be utilized. Until we know more about the conditions on the surface of Venus, we cannot tell what chemical processes occur there. The atmosphere is much denser than that of Earth and has a radically different chemical composition. The pressure at the surface is equal to ninety-one Earth atmospheres. The atmosphere is composed mostly of carbon dioxide, rather than the nitrogen and oxygen that are predominant on Earth. The amount of carbon in the atmosphere of Venus is about as much as exists in all forms on the Earth's crust and oceans. How all this carbon got into the atmosphere on Venus, while it remains in solid compounds on Earth, is an unsolved problem.

There are small amounts of other substances in the air of Venus. Nitrogen is present to the extent of 3 percent in the lower atmosphere, and water to about 0.17 percent, more in total than in Earth's atmosphere. Some of the clouds seem to be droplets of fairly concentrated sulfuric acid. Free oxygen and sulfur dioxide are present, but in very small amounts. As we have stated, the thick atmosphere of Venus absorbs much of the sunlight that hits it. Only 1 percent reaches the surface of Venus. This is in contrast with Earth, where 70 percent of the sunlight reaches the surface. One might think this would make the surface of Venus cold rather than hot, but its temperature depends on a complicated process of exchange of radiation at various wavelengths between atmosphere and surface.

The process by which a thick atmosphere can raise the temperature of the surface is known as the "greenhouse effect," as it was once thought to be the reason why it is warmer than expected inside a greenhouse. The absorption of a lot of radiation by the atmosphere does in fact produce a useful supply of energy for whatever chemical reactions take place there.

It is possible that conditions on Venus were different in the distant past. There may have been an ample supply of free energy available at the surface before the evolution of the present atmosphere. Paradoxically, the lack of a thick atmosphere would also have led to a lower surface temperature because there would be no greenhouse effect. In this case there may have been abundant chemical reactions, of the type assumed to occur in the early Earth, on early Venus as well. If so, there may have been life forms based on carbon chemistry. It is hard to believe that such life could survive the transition to present conditions on the Venerian surface, but this may have happened. Some conditions on Earth today are immensely different from the early Earth, but life has survived both by evolving and by being oblivious to the changes. If the conditions on Venus changed slowly enough, it could be that primitive life there could have evolved to survive and thrive even in the present conditions. Indeed, some of the organisms may have grown to like their present circumstances. After all, even hell is the beloved home for some orders of demons, and present-day Venus, with its heat, storms, and sulfurous clouds, is a fine approximation of hell.

One possibility along these lines is that Venerian life became airborne and survives now in the upper atmosphere, where conditions are more like those of early Venus. The lack of a solid surface need not be a problem for such life any more than it is to the fish that live in our oceans. Achieving buoyancy in the thick atmosphere should not be difficult. This could be done if "airbags" containing gas from the thin upper atmosphere were included in the bodies of the living creatures. An alternate possibility is that the creatures of Venus float on permanent updrafts of wind that blow from the lower to the upper atmosphere. Such floating life forms were proposed by Harold Morowitz and Carl Sagan some years ago. It is unlikely that life could have evolved under these conditions because the density of matter is low, but migration of surface life to an atmospheric home is more plausible. Earthlife has migrated back and forth from water to land and perhaps to air as well; so there is a precedent for such developments on Venus. Future explorations of the atmosphere of Venus should shed light on both of these possibilities.

SOME OBJECTS LESS LIKELY TO HAVE LIFE

A brief walk of forty meters on the +9 level of COSMEL, or about the distance from homeplate to second base on a baseball diamond, brings us to the orbit of the innermost planet, Mercury. The Sun itself is only an additional sixty meters away, a fuzzy yellow ball whose diameter is a bit less than our height. Its gigantic mass and liberal outpouring of energy make it a very likely candidate as a site for life. However, we are concentrating on the cooler, smaller objects of our solar system right now, so we postpone examining it and turn our attention to Mercury. We locate it, a sphere slightly larger than the capital letter O in this text, and revert to the +7 level to examine it in greater detail. Our information on Mercury, based on data from spacecraft flybys in 1974 and 1975, shows it to be a slightly larger version of Moon. It lacks a significant atmosphere and any sign of present or past liquid bodies on its surface. It has an abundance of craters and a wide temperature difference between its day side ($+400°C$) and its night side ($-170°C$, $103°K$). The orbital motion and rotation of Mercury produce days and nights lasting three months. Life does not appear to have evolved on the surface of Mercury. No fluid regions were available in which molecules could interact rapidly. We have not examined the area just below the surface of Mercury, however, or its deep interior. Temperatures a few meters below the surface remain permanently in a range similar to that on Earth, and there may even be liquid water present. Also, this region would be shielded from the intense ultraviolet radiation of the Sun, but enough heat would flow through to provide a periodic source of free energy to the material there. The existence of life inside Mercury remains a possibility, though we have no evidence for it at this time.

We will now return to Earth, having completed our tour of the planets. As we approach our destination, a spectacular sight reminds us that there is yet another type of heavenly object to consider. This is a comet. It seems mainly to be a long "tail" stretching back from a small "head," in the direction pointing away from the Sun. The tail can be millions of kilometers long, longer than our height at the +9 level. In order to view the head up close, we are forced to take COSMEL down to the +4 level. Even at that level, its head (a few kilometers in real diameter) is no bigger than our own. In terms of *mass*, however, the situation is reversed. Almost all of it is contained in the solid head of the comet, not the gaseous tail. The relation of head to tail is like that of a person's body to the breath exhaled on a cold day.

The comet we have seen is one of a group which has an orbit similar to that of a planet but more oval, moving mainly in the region between Venus and Jupiter. Another group has orbits which go far from the Sun, some almost as far as the nearest star, and make a close approach only once in a million years. This second type must be very numerous, since about one a year approaches the Sun in this way. The comets that we are able to observe seem to be composed of a mixture of chemicals very different from those of the Sun or any of the planets. In fact, the mixture is somewhat similar to that found in Earth's biosphere. The elements are in the form of simple molecules, such as water and carbon dioxide. Some scientists have speculated that repeated encounters of comets with the Earth have been the source of many of our volatile elements such as hydrogen and carbon. This may have happened during the formation of Earth, when the encounter rate with comets could have been much higher than now. It certainly has not been the case recently.

We have mentioned the speculations of F. Hoyle and C. Wickramsinghe. They suggest that comets, whose composition is similar to that of Earth's biosphere, have themselves developed life. However, the form of the material in comets poses an obstacle to the evolution of life. In the head, especially in the very concentrated region called the nucleus which contains most of the mass, the material is solid when the comet is far from the Sun and boils away from the surface when it is close. There is not enough gravity on a comet to retain an atmosphere. There is no span of time when a fluid would be present on the surface to permit suitable reactions. Only if pockets of liquid were trapped within the crust would a suitable medium for chemical evolution be available. Even if that were so, a free-energy supply would be a problem. The comet would have to be one that spent most of its time near the Sun. Possibly some energy-bearing molecules could be produced on the surface by absorption of solar radiation and diffuse into the interior.

A final problem is the short lifetime of comets compared with planets. A portion of their mass is evaporated when they approach the Sun. As comets have little chance to replenish themselves, they gradually disintegrate. In summary, comets have some of the features necessary for the development of chemical life, but lack other important ones. They contain a suitable material, and some receive a plentiful free-energy supply on their surfaces. Yet their inappropriate environmental conditions and their instability make comets a long shot for the development of life. There are proposals to investigate a comet by an unmanned space probe before 1990, and if this mission is carried out we may learn enough to decide this question.

Using COSMEL, let us return to the comfort of our homes and the familiarity of a world where our bodies are five to six feet tall. We have seen that the solar system contains an immense variety of environments, most of which are incredibly alien to us and destructive to our type of life. It is very human but also very irrational to assume that ours is the only possible type of life and dismiss the biological possibilities of other environments. When we consider the requirements for life in the broadest possible terms, an entirely different picture emerges. Jupiter may contain most of the life present in the planets as it does the mass.

We will learn the actual state of affairs only by vigorous exploration of our solar system. This will require much more purpose, energy, and time than the imaginary tour we have just completed via COSMEL. We would like to use the results of our COSMEL tour, however, to guide the planning of the real one. With apologies to guidebooks, restaurants, and film reviews, we will award a number of stars to worlds in our solar system with respect to their potential merit as sites for the development of life (Table 10-3).

EXTRASOLAR PLANETS

Of course, many types of planetary environments that might be homes for life do not occur within our solar system. Some are described elsewhere in this book. No worlds are present in our solar system, for example, in which oceans of molten rock or of liquid hydrogen and helium exist. To find such worlds we must travel to other stellar systems. But first we must inquire whether or not planets are associated with most stars.

For a few of the nearby stars, there is some evidence of invisible bodies revolving around the parent star. These invisible objects have masses much less than those of stars; typically, about ten times the mass of the largest of the Sun's planets, Jupiter. In one case, for the star known as Barnard's star, which is among the ten nearest to Earth, a careful observation of its apparent motion over a period of years indicates the possible presence of two or three planets revolving around it, each with a mass approximately that of Jupiter. Other astronomers have disputed the accuracy of these observations. It is not possible with present instruments to observe planets of this size directly. Nor is it possible to detect the gravitational effects of yet smaller Earth-size planets on the motion of their primary stars.

The evidence for extrasolar planets gives some credence to the idea that planets are fairly common objects surrounding stars. Present ideas on the function of planets, based

Table 10-3

A TOURIST GUIDE TO LIFE IN THE SOLAR SYSTEM

Rating	Location	Possible Type of Life
★★★★ (Highest probability; look here first)	Jupiter	Various types at various levels
	Titan (surface)	Life in hydrocarbons or ammonia
	Earth (surface)	Carbon-water life
★★★ (Some information missing; still a good bet)	Saturn	Various types at various levels
	Ganymede (interior)	Carbon-water life
	Callisto (interior)	Carbon-water life
	Titan (interior)	Carbon-water life
★★ (Serious problems known; but some possibilities exist)	Venus (atmosphere)	Floating gas bags
	Mars (surface)	Lichens, superoxide eaters
★ (Unlikely from what we know; but incompletely explored)	Mercury (subsurface)	
	Comets	
? (Little information available)	Earth (interior)	
	Neptune	
	Uranus	
	Venus (surface)	

mostly on our own solar system, are in agreement with this, as they assume that the planets were formed together with the Sun, condensing out of the same interstellar gas. In this theory, the process that led to the formation of the planets was a rapid rotary motion of the protostar that eventually became the Sun. This rotary motion cast out material from the center, which eventually condensed into planets of various types. Such theories of planet formation are in marked contrast to the "catastrophic" theories that were believed earlier in the century. In such speculations planets were formed only after close collisions of stars. This would happen very rarely, at least in our part of the galaxy, and so there would be very few planetary systems. Present theories suggest that planets should be quite common, as they are produced by physical processes that should happen to most stars as they form. Possibly, close double stars are exceptions, since the presence of another massive body nearby might disturb the formation of planets. This would probably not be the case for double stars that are far apart. In fact, one of the stars thought to have a large invisible companion is 61 Cygni, which also is part of a double star separated by ten billion miles. These newer theories are consistent with the reports of several nearby stars with planets.

There have been many discussions of the likelihood that planets of other stars have conditions similar to those on Earth. A detailed treatment of this is given in the book *Planets for Man* by Stephen Dole and Isaac Asimov. This question is certainly interesting if one considers possible human colonization of other planets. The discovery of such worlds would eventually help us answer a question asked earlier: Must evolution on Earthlike worlds inevitably lead to the development of life resembling our own? The discovery of additional worlds resembling Earth will not help us, however, in our understanding of a much broader question: What is the full extent of life in the Universe?

One thing that is relevant to this question concerns the conditions to be expected on extrasolar planets. An im-

Table 10-4

SURFACE TEMPERATURES OF
HYPOTHETICAL PLANETS OF VARIOUS STARS

Type of Star	Star's Surface Temperature (degrees centigrade)	Star's Radius (kilometers)	Planet's Distance from Star (kilometers)	Planet's Surface Temperature
Red dwarf (Proxima Centauri)	2300	2×10^5	(Mercury) 6×10^7	$-167°C$ ($106°K$)
			(Earth) 15×10^7	$-206°C$ ($67°K$)
			(Saturn) 140×10^7	$-251°C$ ($22°K$)
			(Pluto) 600×10^7	$-263°C$ ($10°K$)
Red supergiant (Betelgeuse)	2300	5000×10^5	(Mercury) Inside star	
			(Earth) Inside star	
			(Saturn) 140×10^7	$825°C$
			(Pluto) 600×10^7	$260°C$
Sun	5500	7×10^5	(Mercury) 6×10^7	$170°C$
			(Earth) 15×10^7	$7°C$
			(Saturn) 140×10^7	$-180°C$ ($93°K$)
			(Pluto) 600×10^7	$-229°C$ ($44°K$)
Sirius	9100	13×10^5	(Mercury) 6×10^7	$700°C$
			(Earth) 15×10^7	$345°C$
			(Saturn) 140×10^7	$-70°C$
			(Pluto) 600×10^7	$-175°C$ ($98°K$)
Ultrahot star (Spica)	20,000	70×10^5	(Mercury) 6×10^7	$4460°C$
			(Earth) 15×10^7	$2780°C$
			(Saturn) 140×10^7	$725°C$
			(Pluto) 600×10^7	$210°C$

The planets are assumed to rotate rapidly, to have no atmosphere, and be black spheres. The entry under Sun can be compared with the actual values in Table 10-1.

portant example is the temperature on a planet. This temperature is very sensitive to the amount of radiation that the central star emits, and to the distance of the planet from the star. Some values for the approximate temperature of a planet under various conditions are given in Table 10-4. Because most stars are dimmer than the Sun, the temperatures of many planets will be substantially lower than those on Earth. This, in turn, will affect a planet's surface and atmosphere because for planets of a given size, the temperature is an important factor in determining whether various gases will escape from the atmosphere to outer space, as helium has from the Earth. A question of interest is whether there are planets whose temperature is low enough (lower than 20°K, −253°C) that hydrogen can be liquid or solid on its surface. Such planets would need to receive ten thousand times less radiation from their stars than Earth does from the Sun. The table shows that this should not be an uncommon situation for planets of most stars. On the other hand, planets with surfaces much hotter than Earth can exist circling supergiant stars. However, since these stars have relatively short lifetimes, those conditions will be fairly transient and probably not play an important role in the evolution of life. Even the exploration of very un-Earthlike planets will not provide full information about the extent of life in the Universe. For that, we must examine the Universe itself and the full range of matter and energy it contains. We will attempt this broader search in the next chapters.

Chapter 11

Homes for Life—
The Outside Universe

We have seen that the planets, both those known in our solar system and hypothetical ones that may exist in other solar systems, offer a rich variety of homes for chemical life. Planets, however, comprise only an insignificant portion of the matter in the Universe. In most of the Universe, the conditions of temperature, pressure, and energy flow are such that chemical life, based on molecular combinations and recombinations, cannot exist. For example, in the deep interior of stars, molecules and atoms are broken into ions and electrons. Are such environments necessarily barren of any kind of life? We think that this view is quite parochial. There are various alternatives to chemistry that a life-style could be based upon, both for the storage of order and the source of energy to generate it. The most significant unanswered question about the possibility of such exotic bases for life is whether they can evolve sufficient complexity that we could apply the word life to them. If we did not *know* that Earthlife could have arisen from combinations of carbon, hydrogen, oxygen, and nitrogen atoms, it would be difficult to predict the existence of such an intricate phenomenon. After we have identified some possible homes for life, we will discuss alternative ways in which complexity might evolve.

In searching for life forms that may evolve elsewhere than

on planets, we must first identify other locales which can supply suitable combinations of matter and energy. Our search for possible homes for life will cover the whole Universe. We will again call on COSMEL, so that objects which differ immensely in true size will be viewed on a scale which is convenient to us. We step into the elevator and notice that the largest numbered button is +25. At this level our size will be about one-tenth that of the entire Universe. No +26 level exists in our model. It would be of little use to us to enter a space that we fill almost entirely.

In starting our survey, we will be bold and go to the highest level available. We press the +25 button and step out. For the most part, we see nothing. There is empty space, weakly illuminated by the three-degree microwave radiation that fills the Universe. On looking carefully, however, we see small glows filling space uniformly in all directions, at least to a distance ten times our size. These glows are each about the size of our fingernail, but shine brightly, like a swarm of fireflies lighting the eternal darkness (Plate 18).

The glows are moving like real fireflies, but at speeds comparable to that of light. If we let time pass at its ordinary rate, the speed would seem imperceptibly slow to us at the +25 level, as it would take millions of years for one of the objects to cover the average distance separating it from one of its neighbors. This distance is about three times the size of each object.

The glows are clusters of galaxies. No larger units of matter appear to exist in our Universe. Yet together they all contain only about half the matter that does exist. The remainder is present mostly in the form of isolated hydrogen atoms, spread very thinly in the space between the clusters. To observe one we will aim the COSMEL screen to point into this space, and push the button marked −8. As you will recall, a hydrogen atom on this level is the size of a marble. We are quite lucky and find one hanging in the void of black space immediately outside our elevator door. If we wished to find its nearest neighbor at the −8 level, we

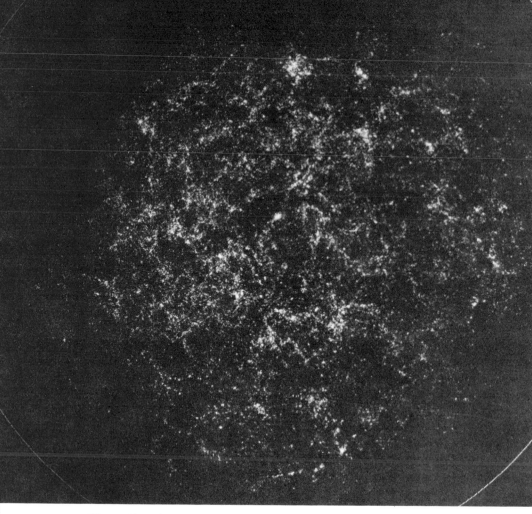

Plate 18. A composite map of the million brightest galaxies visible in the northern sky of Earth. The Universe might appear like this when viewed from the highest levels of COSMEL.

would have to walk a distance equal to that from the Earth to Moon!

There are an enormous number of such isolated hydrogen atoms because space is very large. But there is hardly any way in which they can influence one another, and little energy to help them. Therefore, it is unlikely that the inter-galactic hydrogen can play a significant role in the story of life. We will concentrate instead on the clusters of galaxies.

We return to COSMEL and push the +23 button. At this level we are about the size of a cluster and can observe its parts. As the name suggests, it is made up of individual objects, the galaxies themselves. The cluster is a roughly spherical object, which may contain up to a few hundred galaxies. So the largest object in the Universe, a cluster of galaxies, is about thirty-eight orders of magnitude greater in diameter than the smallest ones, the subatomic particles. Stars lie about halfway between the largest and smallest objects, on the COSMEL levels, while human beings are closer to the very small than to the very large.

At this level the galaxies themselves are mostly oval discs, with the average one similar to a fingernail in size and shape. Their sizes vary within a range of ten, and they also differ in their brightness and form. Some appear as long, wispy spirals, rather than ovals. They all generally have dense centers, diffuse borders, and "haloes" coming out of the plane of the disc (Plate 19). Some galaxies also contain bizarre structures such as long "jets," which may have been ejected by cosmic explosions from the main body. A typical galaxy contains one hundred billion stars, about as many as the atoms contained in a bacterium.

These descriptions may make a galaxy sound disturbingly like some of the living things that we met earlier twenty-five levels down the cosmic elevator shaft, and indeed the structural parallels are eerie. Very probably these resemblances are coincidental, but—? At any event we are not out to bag such big game at this stage, and we will focus our hunt on objects of more familiar size.

COSMIC CLOUDS

We return to COSMEL and descend to level +16 in the outer region of a typical galaxy. Upon emerging we see a large number of wispy clouds of gas and dust, glowing brightly by emitting many forms of radiation. The typical cloud is somewhat larger than we are, and they are spaced a bit farther apart than their individual sizes, giving the appearance of light clouds on a partly overcast night on

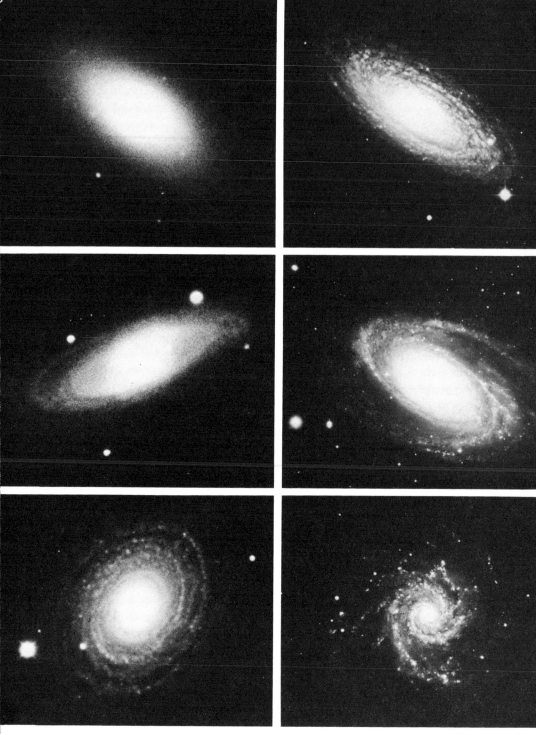

Plate 19. Some typical spiral galaxies, as they would appear on
the +22 level of COSMEL.

Earth. Between the galactic clouds is a more rarefied gas, out of which they have condensed because of the mutual gravitational attraction of their components. Both the gas between the clouds and the clouds themselves are mainly composed of hydrogen and helium, the two simplest of the chemical elements. But there are small amounts of other elements, such as carbon, oxygen, and nitrogen, which have been and are so important in Earthlife.

We can use COSMEL again on its lower levels to examine the composition of a cloud. There are occasional stars within a cloud, produced by a process we will discuss below. These stars play an important role, providing much of the radiant energy that drives physical and chemical processes within the cloud. Especially, the ultraviolet light produced by hot young stars can break atoms up into ions and electrons, which seems to be an essential step in determining how chemical reactions occur.

In order to view the next smallest objects known to exist within the cloud at their own size, we must reduce ourselves to the size of a bacterium at COSMEL level −6. We see solid chunks, called dust grains, of unknown composition. They may be similar to the grains of sand found on a beach. They are widely spaced, and the nearest neighbor to the one we are observing may be thousands of kilometers away at this level. Atoms and molecules also exist within the clouds in the space between the dust grains, and they are also kilometers apart, on the average. This is a very rare collection of matter by Earth standards, but its density is still a million times greater than that of the matter in the space between galaxies. Furthermore, the density varies greatly from place to place in the clouds. In some locations it is as high as that in Earth's atmosphere. But the regions of such high density do not last long. Gravity usually forces them to contract to higher and higher densities, heating up as they contract until they begin to emit radiation because of their high temperatures. It is this radiation that illuminates the cloud, rather than the feeble light emitted by the bulk of the clouds themselves. This process culminates in the formation of stars, which we will soon discuss.

We will pause now to take inventory of the atoms and molecules present in such a dense cloud, conveniently using COSMEL at the −8 level. Helium atoms and hydrogen atoms and molecules, the most common ones in the Universe, predominate here also. We find lesser amounts of heavier atoms such as carbon and, somewhat surprisingly, small organic molecules. We have already discussed the somewhat un-Earthlike nature of these molecules in Chapter 5. It is not at all clear how they were formed, and once formed, how they survive the radiation that tends to break them up. At the densities of atoms in most places in the gas clouds, collisions are very infrequent. It is possible that the interstellar molecules are formed on the dust grains. Atoms may collide with the grains and stick to them as flies stick to flypaper. If many atoms are on the same grain at once, they have a much greater chance of forming a molecule than if the same atoms collide in space. Some surfaces show similar catalytic effects on Earth.

A number of writers feel that the presence of organic molecules in interstellar space is of great significance for the origin of life on Earth. We have discussed this pre-destinist viewpoint in earlier chapters and have explained why we disagree with it. The gas clouds, with their interplay of radiant energy and dispersed molecules, may have a fascinating story of their own to tell once we overcome our egocentric tendencies. In particular, the clouds may furnish the basis for the existence of a totally different form of life—radiant life. We will note that there is a possible biosphere here, and reserve this topic as worthy of treatment in greater depth in the next chapter.

STARS

We next want to study in detail the major components of galaxies, the stars. Stars contain about half the mass of galaxies, and produce most of the radiation by which we see galaxies. Stars are the most characteristic bodies of the present Universe, although this was not true long ago and will not be true in the far future, billions of years from now. In order to appreciate the true glory of the stars, we will first study them where they congregate, in the center of galaxies, rather than in the galactic fringes where our Sun is found.

We will use COSMEL to transport us to a galactic center and then drop to level $+13$. When we emerge, the first impression is of light everywhere. We find ourselves surrounded by stars, many of which shine as bright as the full Moon does on Earth. Within a distance of a hundred meters, we can count thousands of stars of various color and size. If we were at the same level in a part of a galaxy close to Earth, the nearest star would probably be five kilometers away and the total brightness would be that of the Earth sky at night, millions of times less than at the galactic center. Even at the density in the galactic center, the individual stars are too small to see at this level except as points of light. The nearest star, some ten meters away, is relatively as small as a human cell compared with the whole person.

But while the individual stars are too small to see as discs, their total output of radiation is formidable. Although conditions near the galactic center are not very well understood by astronomers, from the estimated density of stars we can guess that the total intensity of radiation at an average place not especially close to any star will be as bright as a cloudy day on Earth. The radiation comes from all directions rather than from one source, so that the general impression would be like the bright sky on Earth (although not blue!). Starlight is a very significant factor in the conditions that exist in the galactic nucleus, whether or not a star is nearby. The suns truly never set at the center of the galaxy, as there are always enough of them visible to provide almost as much light as we have by day on Earth.

Stars contain a large part of the matter in the Universe, especially that part which is at high density and subjected to large flows of energy, both of which are beneficial to the emergence of order. We therefore expect that stars may be one of the main environments where life is to be found. For this reason we will carefully examine some of the important features that may be relevant to the life that inhabits the stars.

In order to begin, we must again decrease our size so that we can examine individual stars more carefully. At COSMEL level +9, we are about the size of an average star. Because the stars are far apart compared with their size, even near the galactic center, we must make sure that we emerge from COSMEL near enough to one to avoid a long walk. To do this we use a locating device in COSMEL to fix on a specific star. We choose as our target our own Sun, whose position is well known. Also, since scientists know the Sun best, what we will describe has the most reliability, although not as much as we would like. Later we will visit other stars that differ from the Sun.

We emerge from COSMEL on level +9 in a region and at a level that we have already explored. It is our own solar system, and we have landed directly in front of the yellow, glowing globe of the Sun. Although when viewed from the distance of Earth or the other planets, it seems solid enough, from close up the Sun is more diffuse, like a meter-wide ball of cotton candy that grows denser as its center is approached. Let us examine it more closely.

THE SUN

We have constructed our model of the Sun in transparent layers so that we can observe the interior readily. This gives us an advantage over experimental scientists, who are limited to direct observation of radiation coming from the Sun's outer layer. Our knowledge of the insides of the Sun and other stars comes from theoretical analysis, and this is somewhat less certain than those scientists responsible for the analysis believe. Nevertheless, it is the best information available, and we have made liberal use of it in constructing our model. This model shows a visible surface, surrounded by a very wispy but hot outer region. This hot region can normally be seen only during a total solar eclipse, when the Sun's visible surface, called the photosphere, is obscured by Moon. The photosphere is by no means solid. It only appears so because all the visible light from the Sun originates from a thin sphere, which our eyes interpret as a solid body. We have seen the same is true for Jupiter and some of the other planets. There is no sharp change in density at the photosphere corresponding to the difference in density between the atmosphere and the surface of a solid planet like Earth. The density of matter increases slowly inward from the photosphere and decreases slowly outward. The density of the part of the Sun which is above the photosphere is about a thousand times less than that of the atmosphere near the surface of the Earth. The temperature at this point in the Sun is about 4000° C. We can be grateful again that we are only dealing with a model. About halfway down through the photosphere the temperature is 5700° C, which is usually taken as the temperature of the Sun's surface. As we go up outside the photosphere, the density drops but the temperature rises up to a million degrees or more. However, so few atoms are present at these high temperatures that this region is essentially a hot vacuum.

The atoms present in the photosphere, and presumably the interior of the Sun as well, are mostly hydrogen and helium in a ratio of about ten to one. This is approximately the same as that found in the gas between the stars, and represents the average situation in the Universe. In addition, the Sun contains about 2 percent, by mass, of heavier atoms such as carbon, magnesium, iron, and silicon.

As we proceed inward from the photosphere of the Sun, there is a gradual increase in pressure, temperature, and density. The rate of increase with depth is not so different from that in the interior of the Earth or in a planetary atmosphere, but the distances are so much vaster that the values of pressure and temperature eventually reached are immensely greater than anything possible in a planet. Because of the gradualness of the change, there is no solid surface anyplace. The Sun is basically one large atmosphere, in spite of its solid appearance. To emphasize our search for habitats for life we have divided the Sun into three zones. We will visit each zone (using the −12 level of COSMEL) to examine the conditions and state of the matter there. (See Fig. 27.)

Our first zone is quite narrow, no more than the thickness of an apple's skin compared with the apple itself. The site we visit is about one millimeter into the interior of our 1.2-meter globe. The temperature is 30,000° C. At this temperature the atoms have lost some of their electrons, which wander about freely like small children that have temporarily escaped from the clutches of their overfond parents. The overall density is still low here, no greater than that of the air at the surface of the Earth. But because of the high temperature, the pressure this gas exerts is very high, equal to that deep under one of our oceans.

The next site we visit is halfway toward the center of the Sun. Here the temperature is several million degrees, and the gas has been compressed to the density of water. Matter here does not, however, behave like any of our familiar substances on Earth. The electrons have been more completely detached from their atoms and wander about freely, leaving the nuclei behind as positively charged ions. The separated positive and negative electric charges move independently of each other. This state of matter is called a plasma. We consider this zone to be a very promising one for the development of life. We will mark it as a possible biosphere, and reserve it for a more complete discussion in the next chapter.

Our last stop is at the very center of the Sun. The temperature here reaches ten million degrees, and the density is five times greater than that of solid gold, greater than any material known on Earth. The matter does not behave

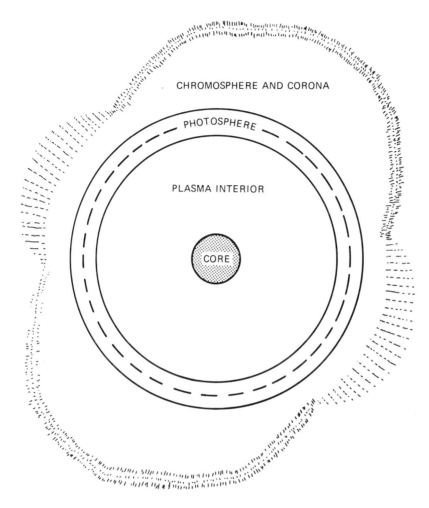

Figure 27. The interior of the Sun shows a gradual change in conditions with distance from the center, rather than abrupt transitions. The central core, in which nuclear-fusion reactions occur, has the highest temperatures and densities. A large region containing plasma surrounds this core. The most readily visible part of the Sun, the photosphere, lies outside of the plasma region. Still farther out from the center is the extremely tenuous corona.

at all like a solid, however, but is again in plasma form, in which separated positive and negative charges move independently of each other.

The high temperature and density near the center of a

star allow the star to produce the energy that eventually escapes the surface as radiation. The heat in the center of the Sun originally came from the potential energy that was lost when the Sun was formed from a dense gas by the gravitational attraction of the matter in the gas on itself. This process is similar to the one which takes place when the water in a waterfall hits the bottom and becomes warmer. A star becomes much hotter than a waterfall because the matter in it has so much farther to "fall." Few stars are being formed by this process at any time compared with the total number visible. The birth and growth to maturity of a star like the Sun takes only a few million years, compared with the billions of years of its lifetime. Most stars, including the Sun, are adults, and there are very few stellar "children" to be seen.

The amount of energy that the Sun obtained by "falling together" at its birth was sufficient only to keep it shining for about thirty million years, not nearly enough time for the known origin and evolution of Earthlife to have taken place. The Sun and other stars have kept shining for billions of years only because they found a new energy source, nuclear fusion. The nuclei of hydrogen atoms combine to produce helium together with radiant energy, and the outward pressures of the hot gas and radiation balance the inward force of gravity and keep the Sun from contracting further. This fusion process thereafter provides a reliable energy source for the Sun.

SOME OTHER STARS

Now that we have some idea of what one star looks like close up, we can take a trip to some other stars in our galaxy to see what similarities and differences they show. In order to make this voyage, we must return to some of the higher levels of COSMEL, since the $+9$ level at which we examined the Sun, the distance to the nearest star is equivalent to walking around the Earth on level 0. A convenient level for traveling between stars in the galactic boondocks where the Sun exists is +16. We emerge from COSMEL and see the familiar night sky of Earth, except that both the stars of the northern and southern skies of our planet are visible. However, these individual stars can be seen only as minute points of light at the +16 level. Each has been reduced to the approximate size of a virus on Earth. The orbit of Pluto around our own Sun is about as large as a dot that is barely visible.

We find that there are about twenty stars within ten meters of our Sun, and that each star is approximately ten meters from its neighbors (one meter now corresponds to one light year, the distance that light travels in one Earth year). We will approach a number of these neighbors of our Sun on level +16, then return to the +9 level (the size of our own Sun) and inspect each in turn.

A list of the twenty stars nearest to the Sun, with a few of their properties, is given in Table 11-1. Of these stars only two are larger than our Sun, the bright stars visible in our winter skies as Sirius and Procyon. A third star visible only in the southern hemisphere, known as Alpha Centauri, is about the same size as the Sun, while all the other nearby stars are smaller than the Sun, both in radius and in the amount of matter they contain.

Most of the stars in the vicinity of the Sun are similar in composition; that is, they contain mostly hydrogen and helium. They also contain small amounts of the nuclei of heavier atoms, such as carbon, oxygen, and iron.

Like human beings, stars may be found that "live" alone or with one or more companions. Sirius, for example, has a small companion star which we will inspect closely in a little while. Half of the one hundred stars nearest to Earth are members of such multiple star systems, and a similar proportion is believed to hold elsewhere. The distances between the stars in a "family" may vary from values barely greater than their diameters to much greater separations. Where the stars are very close, their shapes may be distorted

Table 11-1

THE NEAREST STARS TO THE SUN

Star	Distance (light years)	Surface Temperature (degrees K)	Mass of Star Divided by Mass of Sun	Special Features
Proxima Centauri	4.2	2600	.1	Triple Star System
Alpha Centauri A	4.3	5700	1.09	
Alpha Centauri B	4.3	4100	.88	
Barnard's Star	6.0	2600	~1	
Wolf 359	7.7	2500	~1	
Lalande 21185	8.2	3000	.35	
Sirius A	8.7	9400	2.31	Double star
Sirius B	8.7	8000	.98	White dwarf
Luyten 726-8A	8.7	2600	.044	Double star
Luyten 726-8B	8.7	?	.035	
Ross 154	9.3	2600	<1	
Ross 248	10.3	2400	<1	
Epsilon Eridani	10.9	4500	~1	
Ross 128	10.9	2600	<1	
Luyten 789.6	11.0	2500	<1	
61 Cygni A	11.2	4000	.59	Double star
61 Cygni B	11.2	3700	.50	
Procyon A	11.4	6500	1.75	Double star
Procyon B	11.4	6000	.64	White dwarf
Epsilon Indi	11.4	4000	~1	

The names of all the stars depend on the catalog in which they are listed. Note the predominance of small, relatively cool stars.

because of gravity, and there may be a flow of material from one to the other.

Let us now wander a bit farther afield into the galaxy and describe the interesting types of stars we encounter. We certainly have enough room in which to roam. The galaxy is one hundred kilometers in diameter at the +16 level! As we examine a wide range of stars, we find that our Sun is larger than average. As a class, stars vary in mass by a factor of one thousand, from one-tenth that of the Sun to one hundred times greater. This is a wider size range than that of adult human beings, but a much smaller one than that of mammals, which vary from shrew to whale by a factor of one hundred million. Smaller stars than the ones noted above may also exist, unobserved by us because they emit too little radiation.

Apart from their size, stars differ in their color and brightness. When we observe one, we find that there is a fairly simple relation between its color, surface temperature, and the amount and type of radiation that it emits. A yellowish star like our Sun has a surface temperature of about 6000° C, hot enough to boil all substances. Such stars mainly emit the radiation we call visible light and much smaller amounts of infrared and ultraviolet radiation. A star that is bluer or whiter than our Sun is also hotter, perhaps up to 30,000° C. These stars emit much more radiation per unit of surface area than the Sun. Furthermore, one type of white star is much larger than the Sun, so that the total radiation emitted may be as much as a million times greater. Also, this star emits proportionately more ultraviolet radiation and less infrared and visible light. These prodigal suns tend to have short lives because the radiation they emit so copiously derives from a profligate use of nuclear fuel. This fuel is used up in a few million years, a short time compared with the many billion years that our Sun will be able to shine. One consequence of this state of affairs is that such bright stars are rare in the present Universe, and the ones we see have all been made in the comparatively recent past. Some have even been formed after human beings already roamed the African plains. The ultimate fate of these and other stars after they have used up their nuclear fuel will be discussed shortly.

Much more common than the brightest stars are small reddish stars whose temperature may be as low (!) as 2500° C, at which value some metals remain solid. Going closer to such a star, we find that it emits less than 1 percent as much radiation as the Sun, and that radiation is mainly infrared, with very little visible light. These stars are very common and constitute perhaps 90 percent of all stars at present. They are also incredibly long-lived because they emit so little radiation. In the words of astronomer J. L. Greenstein, "In a hundred billion years, the Sun will have faded into a dead, small, infrared, degenerate star probably invisible from the Earth." Most presently existing stars will have burnt out. But a red dwarf star "will still have its full youthful vigor with ninety-five percent of its long life ahead." Truly, the meek shall inherit the Universe. Perhaps someday if we have not found a better solution in the interim, human society will transplant itself to the vicinity of a red dwarf star, extending our future for billions of years.

If life in the vicinity of these stars develops at anything like the same rate it has near the Sun, it will have a very long and probably glorious future ahead of it. However, planets circling such stars at a distance may have temperatures lower than those encountered on planets in our own solar system, with seas of liquid hydrogen covering them. Such an environment would preclude any chemical basis for life, but would open other opportunities for it.

If we are willing to take a longer walk on level +16, we can visit a rather different type of star than the ones in the vicinity of our Sun. A hike of four hundred meters (four hundred light years at ground level) brings us to the star Antares, visible from Earth only as a red point of light. It is still a point of light when viewed on level +16, but if we went down to level +9, where the Sun was about a meter in diameter, Antares would be a giant ball, some five hundred meters across. At this size, the orbits of all the planets of our solar system out to Mars would fit inside Antares. Some even larger stars have been discovered, and very recently, by using subtle computer methods for reconstructing images, it has actually been possible to photograph one red star (Betelgeuse) as a disc rather than as a point of light (Plate 20). Stars of this type are known as red supergiants, for obvious reasons.

The outermost layers of the atmosphere of such stars are no denser than what we would consider to be a good vacuum on Earth, although they are still much denser than the interstellar space material. Even deep inside the atmosphere, 90 percent of the way toward the center of Antares, the density is perhaps only 1 percent that of the atmosphere at Earth's surface. The temperature at this point is several thousand degrees centigrade. Some molecules can exist under these conditions, and indeed have been identified by the characteristic radiation they emit. The numbers of molecules present are immense by Earth standards. The number of water molecules in the atmosphere of Antares, for example, might be a million times greater than the amount present in Earth's oceans. Furthermore, the densities and temperatures in these stellar atmospheres are high enough that collisions between molecules are much more frequent than among the molecules of interstellar space, even in the dense clouds. It is possible that a rich, high-temperature chemistry could develop in the atmospheres of supergiant stars and serve as the basis for a form of life.

When we proceed to the innermost core of Antares, a region whose size is not much larger than the Earth's, we encounter much denser material. In fact, its density is greater than that of any known solid and up to a million times as much as that of water. If we use COSMEL at its atomic level (—8) to look into the core of a supergiant such as Antares, we find that many of the elements heavier than helium that are found in the Universe are being made there by nuclear

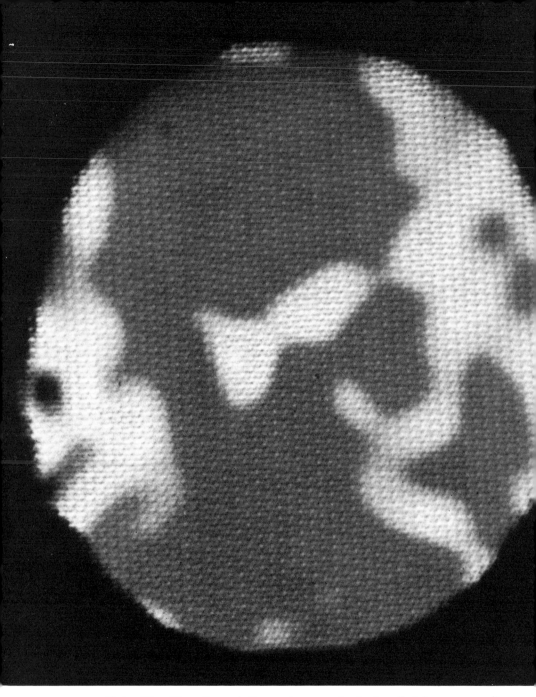

Plate 20. The disc of the red supergiant star Betelgeuse, as it might appear on COSMEL level +13. The photograph was taken at the Kitt Peak Observatory, using a technique known as speckle interferometry. The origin of the structure shown on the star's surface in the photograph is unknown.

fusion reactions. As we will see, some of these stars later explode, spewing their heavy-element contents into interstellar space, where they mix with hydrogen and helium. An enriched gas of this type was involved in the creation of our own solar system. Since Earthlife depends on substantial amounts of heavier nuclei such as carbon being present, we owe our existence to the processes in these dense, hot cores.

Clearly something strange has happened to convert a normal star to a supergiant like Antares. What has happened is not completely understood, but it is believed that very massive stars become supergiants at a certain stage in their lives. When most of the hydrogen that provides the nuclear fuel for the star has been used up, the central core of the star is no longer supported by the pressure of its energy release against the tendency to contract under its own gravity. The core therefore does contract to a high density and temperature until at a temperature of one hundred million degrees, the helium in the core begins to undergo nuclear fusion to produce still heavier elements and more energy. This new release of energy in the form of radiation "blows out" the outer regions of the star into a thin atmosphere, and the supergiant is formed. Woe to any planets that happen to be within the region into which the star's atmosphere expands in this process. They become heated to the temperature of the stellar atmosphere, which may be 3000° C. However, the process takes a long time, perhaps thousands of years, since the gas in the atmosphere is so thin.

STAR'S END

All sources of free energy must come to an end, including the nuclear fuel in stars. When all the light elements like hydrogen and helium are converted into heavy elements like iron, no more nuclear energy can be extracted. The stellar core has been kept from collapsing only by this energy generation, somewhat as a balloon does not collapse as long as air is blown into it. But now the gravitational force seeking to pull the core together reasserts itself, and the central core contracts still further. There are several possible end results to this process, depending on the mass of the star.

The most common result, which happens to most stars less massive than the Sun, is the formation of a starlike object called a white dwarf. It is estimated that there are some ten billion such stars in our galaxy. We can visit this type of star by retracing our steps and returning to the vicinity of our Sun. When we revisit the bright star Sirius, we recall that it has a nearby companion star (Sirius B), which is a white dwarf. On descending to the +9 level and approaching it, we find that Sirius B is very small, only about 50 percent larger in radius than the Earth, but that small package contains as much mass as our Sun. Sirius B has a high surface temperature, but emits much less radiation than the Sun because it is so small. Entering it, we find that while there is more hydrogen and helium in the outer layers, the inner regions comprise mostly heavier elements like carbon, oxygen, and iron. Furthermore, there are no atoms inside the white dwarf. The electrons are separated from the nuclei by the high pressure and roam freely. In fact, this allows the white-dwarf core to resist the tendency of gravity to contract it still further. Electrons do not like being compressed into a small volume of space, and this reluctance, known to physicists as Pauli's exclusion principle, produces a counterpressure to gravity which stabilizes the star at a radius about that of the Earth. The nuclei in the white dwarf, being much heavier than the electrons, form a solid or liquid similar to those familiar to us but a million times denser. The arrangement of nuclei in the solid will be complex if many different types of nuclei are present inside the star.

Because most stars are low in mass, they will become white dwarfs. Eventually, they will radiate away all their internal energy and end up as "black dwarfs." However, this process will take billions of years because in the late stages the star is cool and radiates very little energy. So a white dwarf lasts a very long time and has a complex inner structure, immense density, and after a while, rather low temperatures measured in hundreds or thousands of degrees. For much of its lifetime, there will be sizable energy flows from the interior of the star to the outside. All these assumptions point to the possible development of complex forms of life within the star. An ordering of the positions of the nuclei in the solid parts of the interior is an obvious possibility for the basis of one form of life. But in view of the uncertainty among scientists about the detailed conditions inside white dwarfs, we shall not speculate further here about what life may have developed there. There may be some white dwarfs that have already cooled down to a temperature so low that they have become almost invisible "black dwarfs." The surface temperature might then be similar to that of the Earth, at least for many millions of years. Since the object is fairly rich in heavier elements, it will eventually be a planetary body with a mass like that of a star. Such bodies may be fairly common in the Universe, and will become more common in the future as more white dwarfs cool toward oblivion. There might be a flow of energy from the warmer interior of the star, and possibly hot spots would develop because of an uneven distribution of matter inside the star. Under these conditions, the development of chemical life is not hard to imagine. One significant difference from planets would be the immense gravity, almost a million times greater than that at Earth's surface. This would tend to make life and all other phenomena on the stellar surfaces very two-dimensional. Things that are typically kilometers high or deep on Earth would tend to be millimeters on these stars. But this in itself does not preclude the existence of life related to that on Earth with a somewhat different geometrical arrangement. Also, floating objects of bacterial size would be shielded from the effects of gravity. If there are numbers of black dwarfs in the Universe, yet another possible home for rather familiar life may exist.

If a star is somewhat heavier than the Sun but not more

than about twice as heavy, even the resistance of electrons to being squeezed together is not enough to keep gravity from compressing the star even more. The electrons are jammed into the positive nuclei, and the two combine, transforming the protons in the nuclei into chargeless neutrons. This process does not leave the star unscathed. A huge burst of energy is released, much of it in the form of peculiar subatomic particles called neutrinos. This burst of neutrinos, accompanied by some electromagnetic radiation that we can easily detect, is observed as a rare type of stellar explosion known as a supernova.

What is left behind after the supernova is most of the star's mass, now converted to neutrons and held together by gravity. The neutrons also resist further compression to some extent, and if the remaining star is not too heavy, a stable object results known as a neutron star. We can visit a neutron star by taking a trip with COSMEL to a diffuse object visible from Earth called the Crab nebula (Plate 21). It is about five thousand light years away, or about a five-kilometer trip on level +16. The stars are beautiful, and it is a lovely night for a stroll, so we hike over to it. A supernova at the position of the Crab was visible to Chinese astronomers in 1054 A.D. Many years later it was found that a faint star within the Crab nebula was emitting regular radiation pulses of various wavelengths, radio waves, light, and X-rays. This "pulsar" was eventually understood to be a rotating neutron star, the remnant of the supernova. There may be millions of such neutron stars within our galaxy.

When we attempt to view the neutron star in the Crab nebula from close up by descending to level +9 as we usually do, we are in for a surprise. While our own Sun was a four-foot globe at that level, the neutron star remains a point of light. We have to retreat all the way down to level +4 to make it readily visible. It is a metallic sphere whose diameter is the size of a human being. It appears to have a crust no thicker than our small finger and a much denser interior. In our normal world it is only about twenty kilome-

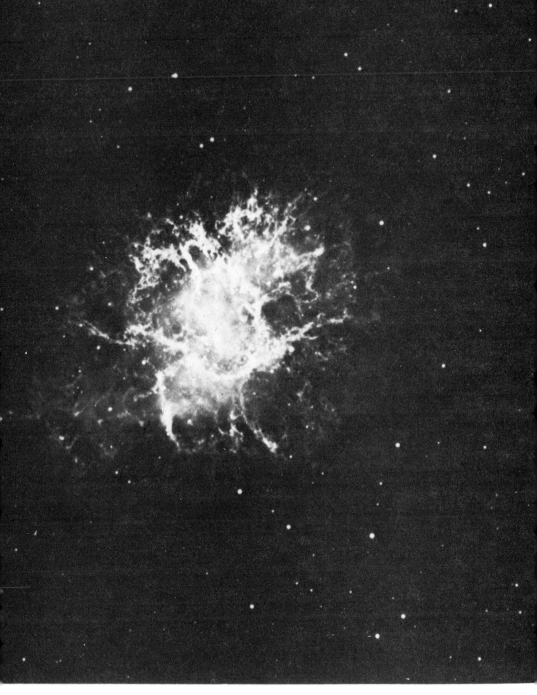

Plate 21. The Crab Nebula, as it might appear on the +18 level of COSMEL. A faint object within the nebula is the neutron star remnant of a supernova explosion that was visible on Earth in 1054.

ters across, incredibly tiny for a star. Its size relationship to a supergiant is roughly that of a bacterium to a whale. In order to push all a star's mass into a sphere that size, the density must be a hundred trillion times greater than that of water, or a hundred million times greater even than white-dwarf matter. An object the size of a bacterium made of material this dense would weigh about a kilogram. Basically, a neutron star is like a giant nucleus of an atom, held together by gravity and extending over a distance immensely greater than the size of an ordinary nucleus.

Because of the very high density, the gravity outside a neutron star is very high, billions of times stronger than that on Earth's surface. This gravity would tend to squash any ordinary object flat against the surface of the star, down to a layer only a few atoms thick. But this will not bother us on our imaginary journey.

As we enter the neutron star, we find that its outer layer has an intricate physical structure. In this outer layer the pressures are not great enough to force the electrons into the nuclei; instead there is matter similar to that in a white-dwarf star, with separate electrons and nuclei. In all of this region, except for a surface layer that is only a few meters thick in actual size, the nuclei form a solid network and the electrons form a gas that flows between the spaces of the solid. This crust is a kilometer or so deep in actual size. Although a million times denser than water, it is so much less so than the inner star that it is really more of an atmosphere than a crust. This part of the star contains a variety of types of nuclei, typically those with intermediate atomic weights such as iron. The matter is not homogeneous in the crust, and it is even possible that "chemical compounds," in the form of solids containing several nuclei in a regular structure, occur.

There are other effects on the matter in the thin surface layer of the neutron star that drastically alter its properties. These are magnetic forces in and outside the neutron star that are much stronger than any that exist or have been produced on Earth. Under the influence of such strong magnetic forces, atoms are distorted from their normal shapes into weird configurations (Fig. 28). These "magnetic atoms,"

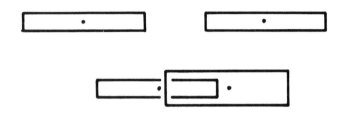

Figure 28. Magnetic atoms. Under the action of the immense magnetic forces present on the surface of a neutron star, the shapes of atoms become distorted into long, thin forms such as the hydrogen atoms pictured here. Such atoms can also fit into one another to give analogues to molecules.

whose existence was proposed by the physicist Malvin Ruderman in 1970, can be much more tightly bound than normal atoms. Also, they bind with each other to form molecules in a very different way. The typical form of such molecules is a long polymer-like chain of atoms, in which the atomic nuclei line up along the direction of the magnetic force and the electrons form a kind of sheath surrounding the line of nuclei (Fig. 29). Two such chains will be attracted to each other by electrical forces (Fig. 30). They are also quite strong, much more so than steel or any ordinary materials. They are even strong enough to stand perpendicular to the surface of the neutron star and can resist its immense gravitational pull, whereas ordinary materials would crumble into rubble a few atoms thick.

It is not well understood to what extent the properties of the chains, or the interactions between chains, depend on the nuclei present in them. But there seem to be ample possibilities for the creation and storage of order in the arrangement of atoms along the chain. There is also a possibility of a replication of order by the formation of new chains in paral-

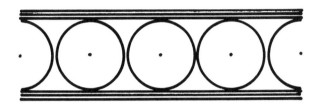

Figure 29. Magnetic atom polymers. An array of magnetic hydrogen atoms can fit together into a long polymerlike chain, in which the nuclei lie along a central line and the electrons surround this line as pictured in (a). For heavy atoms such as iron, some of the inner electrons will remain in roughly circular orbits, while the outer electrons will be distorted into long, thin shapes, and the polymer will appear as in (b).

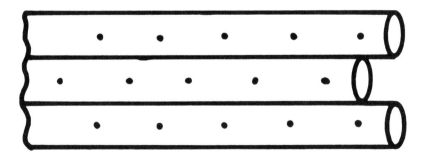

Figure 30. Binding of magnetic atom polymers. Nearby polymers will attract one another, and will bind together into a pattern in which the nuclei of adjacent polymers are displaced laterally, as shown in the figure.

372

lel with existing chains. This requires a source of free energy to break apart old chains, or a supply of new atoms coming off the surface. Both of these are available from outside the neutron star. A rain of energetic ions and electrons should constantly be hitting parts of the surface because of the influence of the same magnetic forces that bind Ruderman's magnetic chains. These forces can trap charged particles into orbits near the star and cause many of the particles to collide with the surface. The energy flow onto the surface of the neutron star can be larger than the energy flow out of a star like the Sun, and it is concentrated onto a very small space so that it is quite intense. These energetic charged particles can easily eject a nucleus from a chain if they hit one. This should happen about once per second for a typical nucleus on the surface of the star. Therefore, an ample supply of the building blocks for new chains should be available. As long as strong magnetic forces are associated with neutron stars, the surfaces will contain large amounts of exotic, dense forms of matter. With an energy flow available from outside, complex forms and structures may arise. Once again, we have a possible home for life.

Under the crust of the neutron star, the matter is almost pure neutrons with a 1 percent mixture of protons and electrons. It is thought that these objects form a kind of fluid, with no rigid structure extending over large distances. The fluid may have remarkable properties similar to those of liquid helium at very low temperatures. Such a "superfluid" extending over many kilometers can have a variety of types of internal motion with a potential for becoming orderly. In this case the rotational energy of the whole neutron star would provide a source of free energy to help the ordering process. Physicists are just beginning to understand some of the phenomena going on inside of a neutron star, and we may expect that as soon as these are understood, some suggestions will emerge as to the types of complex order that can be present. At that time the question of life at the highest densities we know of can be examined in some detail.

Table 11-2

SOME ENVIRONMENTS OF POSSIBLE INTEREST FOR LIFE

Environment	How Common It Is	What Is Found There	Types of Energy Flow
Interstellar clouds	Thousands in each galaxy. One cloud may contain a mass of a billion Earths	Hydrogen, helium, and some heavier atoms; many types of molecules; and dust grains. All are present at very low densities	Visible and ultraviolet light. Cosmic radiation
Atmospheres of giant stars	Billions in each galaxy. Each atmosphere contains more material than the whole Earth	Hydrogen, helium, and some heavy atoms. Some types of molecules. Densities are similar to Earth's atmosphere	Intense radiation from star's interior. The temperature is several thousand degrees centigrade
Solid interiors of planets	Perhaps 10^{12} in each galaxy	Heavy atoms and molecules in liquid or solid form. The density is slightly greater than Earth-surface matter	Heat from radioactive decays

Interiors of white-dwarf stars	Billions in each galaxy. Each white dwarf contains as much material as 100,000 Earths	Heavy nuclei such as carbon in a liquid or solid arrangement. Electrons float in the liquid. Densities are millions of times greater than Earth matter	Some radiation and heat flow from star's center
Solid neutron star crust	Perhaps 100 million in each galaxy. Each crust contains as much mass as the Earth	Heavy nuclei and electrons. Very intense magnetic fields. Atoms may polymerize in intense fields. Densities are like the interior of white-dwarf stars	High-energy particles from outside. Radiation from interior
Liquid interior of neutron star	Perhaps 100 million in each galaxy. Each star contains as much mass as the Sun	Neutrons, some protons, and electrons. Densities are trillions of times greater than matter at Earth's surface	Some radiation from inside. High stored rotational kinetic energy

In our survey of the end results of stellar evolution, we have not yet discussed what happens to a star much more massive than the Sun after it uses up its supply of fusible nuclei. As far as we know now, such a star will continue to contract indefinitely even after it is composed of all neutrons, and will approach closer and closer to being a black hole. This is not the place to give a description of the properties of black holes. The most relevant one for our purposes is that no information about what happens inside a black hole can be obtained except by someone prepared to take a one-way trip inside to see. In this respect the outside and inside of a black hole are related as are the present and future. Because of this restriction, it appears idle to discuss life inside a black hole originating from stellar collapse, as we are unlikely to learn anything about it without a price that we are unwilling to pay. Enough environments which may be hospitable to life exist in the accessible parts of the Universe that we need not greatly lament omitting one from our further consideration.

It is time to return home from our tour of the Universe. We will also return in our minds to the vicinity of our own Sun, then to our own planet, to our living rooms, and the level with which we are quite familiar. We have seen a number of habitats that may be homes for more exotic and strange life forms than the betentacled monsters constantly offered us in science fiction. We will discuss some of the more exciting of these forms in the next chapter. Meanwhile, we have summarized a few of the likely environments for life in the Universe in Tables 11-2 and 11-3, just as we did in Table 10-3 in Chapter 10 for life in the solar system.

Table 11-3

A TOURIST GUIDE TO
LIFE BEYOND THE SOLAR SYSTEM

Rating	Location	Possible Form of Life
★★★★		
	High interiors of ordinary stars	Plasma life
	Interiors of white-dwarf stars	High-density life
★★★		
	Interstellar gas clouds	Radiant life
	Atmospheres of giant stars	High-temperature chemical life
★★		
	Neutron star surfaces	Magnetic-atom polymer life
	Very cold planet surfaces	Solid-hydrogen life
	Black-dwarf surfaces	Chemical life
★		
	Core of ordinary stars	?
	Typical interstellar space	?

The ratings are on the same scale as in Table 10-3. Plasma life, radiant life, and solid-hydrogen life are discussed in greater detail in Chapter 12.

Chapter 12

Physical Life

We have seen that in many of the environments found in the Universe, conditions are so different from those we know on Earth that even the forms of matter present there are unfamiliar. Yet the matter in these environments may meet the requirements necessary for life. In Chapter 11 we identified a number of locations where life based on other physical effects than molecular combinations may have developed. We will now explore in more detail the features of three possible forms of such physical life. The habitats of these speculative physical life forms vary enormously in their properties, yet they are all fairly common in the Universe. Our knowledge of these environments is almost entirely theoretical at present, so that in trying to imagine how they could evolve the complexity necessary for life, we are really extrapolating far beyond our basis in observation. Scientists have been able to do this successfully in some cases, such as understanding the very early history of the Universe, but it is never easy and always uncertain. These limitations should be kept in mind as we continue our discussions.

Our descriptions of physical life will emphasize the alternate methods by which order is stored, since this will be the greatest difference from Earthlife. The sources of free energy already cover a wide range in the forms of life we know, so variations here are less surprising, even though the range will be much greater in extraterrestrial life. We will also make some very speculative remarks about the directions in which various types of physical life might develop and the types of organisms that might result.

LIFE IN THE SUN

> *So strong has been the belief that the Sun cannot be a habitable world that a scientific gentleman was pronounced by his medical attendant to be insane, because he had sent a paper to the Royal Society in which he maintained "that the light of the Sun proceeds from a dense and universal aurora which may afford ample light to the inhabitants of the surface beneath, and yet be at such a distance aloft, as not to annoy them."*

Thus wrote Sir David Brewster in 1854 in describing the fate of a certain Dr. Elliot in 1787. Sir David, of course, did not agree with the diagnosis. He was engaged in a debate on the question of extraterrestrial life with Dr. William Whewell, and he was advocating the extreme case—that every globe that had been created was inhabited. To most observers at that time, a world needed to be like the Earth to be habitable. Dr. Elliot had affirmed that "vegetation may be obtained there as well as with us"; that "there may be water and dry land there, hills and dales, rain and fair weather," and that "as the light and the seasons must be eternal, the Sun may easily be conceived to be by far the most blissful habitation of the whole system."

It has been recognized, of course, since ancient times that the Sun, as the source of light and warmth for the Earth, was indispensable for life on this planet. Such diverse peoples as the ancient Greeks and Egyptians and Amerindian tribes considered it to be alive in itself and gave it divine status. As scientific knowledge increased about the nature and temperature of the Sun, these views became uncommon. Thoughts of the type voiced by Dr. Elliot represented a last-ditch attempt to rescue at least some area of the Sun as a possible home for life. These ideas were given much greater respectability when they were advocated by the celebrated astronomer Sir William Herschel. He had devised improved telescopes, made fundamental discoveries about the nature of stars, and discovered the planet Uranus.

Alas, even so noted an astronomer can be wrong. We have described the modern view of the Sun in the previous chapter. An environment with a temperature of from thousands to millions of degrees, in which matter has been converted to plasma form, would hardly be considered to be blissful for Earthlife. A modern scientist who advocated life within the Sun would be looked on as eccentric by his colleagues. And yet Sir William Herschel may, in part, have the last word. Let us quote Sir David Brewster again: "Sir William has never asserted and never did believe that the children of the Sun were to be human beings, but, on the contrary, 'creatures fitted to their condition as well as we on this globe are to ours.' "

In an earlier chapter we marked a portion of the interior of the Sun as a biosphere of promise. What type of creature would be able to dwell in this particular condition?

PLASMA LIFE

Our solar biosphere is an area below the photosphere, yet somewhat removed from the center. This is the home of the "plasmobes." It may be difficult to visualize these beings and the realm they inhabit—Plasmaland. There are no permanent spatial structures there, nor even any molecules present. Matter is in the form of a plasma and consists of positively charged ions and detached electrons, both acted on by intense magnetic forces. (Such forces have been detected on the surfaces of stars; for example, in what we know as sunspots. They probably also exist within the Sun with much greater intensity, although we cannot observe them directly.) While such an environment may seem a strange habitat, we think that it can nevertheless serve as a suitable arena for the evolution of ordered structures, of life, and even of individual living creatures. The plasmobes are composed of patterns of magnetic force, together with groups of moving charges, in a kind of symbiosis. The possibility of living patterns of magnetic force within stars was suggested some time ago by the physicist A. D. Maude, who proposed a method by which certain patterns that occur by chance replicate themselves. In Maude's model the star gradually accumulates a large number of copies of a few such replicated patterns. This scenario is somewhat similar to the one in which Earthlife evolved after the chance production of a replicating molecule. The plausibility of this type of replication happening inside a star would depend on how complex a magnetic pattern would have to be, in order to replicate. If this can occur for very simple patterns, as Maude suggests, then life in a star might indeed arise without any need for a long period of prebiotic evolution.

The hypothetical inhabitants of Plasmaland have a more complex basis for their life, involving charges as well as magnetic forces. Their negative electrons and positive ions interact through collisions as they do in some liquids under Earth conditions. In addition, they respond to the presence of magnetic forces by modifying the way that they move, sometimes by forming arrangements of moving charges that are stable against disturbances from the outside. It is through the use of magnetic forces to produce such stable patterns of motion that Earth scientists are attempting to harness nuclear fusion for the production of useful power.

The effect of magnetic forces on charges is reciprocal, in that moving charges also influence the pattern of magnetic forces, which in turn further influences the motion of the charges. This situation, vaguely analogous to the reciprocal influence of proteins and nucleic acids in Earthlife, can lead to a more rapid growth of order than a situation in which an unvarying quality affects a mutable one.

The development of life in Plasmaland may have started with random motion patterns of the charges. Some of these patterns proved fruitful. The magnetic forces they produced in turn stabilized the patterns of motion that created them. It was a "you scratch my back and I'll scratch yours" situation. It has been found both in experiments on Earth and in calculations that such self-stabilizing arrangements of plasma can occur. Other random patterns produced magnetic forces that dissipated the charges. The frutiful arrangements persisted and, by a kind of natural selection, eventually became the dominant type. Part of the inside of the star became orderly, in the sense that only a few of the patterns of moving charges that were possible did occur, in obvious analogy to the occurrence of only a few of the many possible carbon compounds on Earth. These favored arrangements of charge, through their magnetic forces, acted on other random arrangements to convert them to the favored form. Since this represented an increase in order, a supply of free energy was needed.

The most likely source of free energy for a solar biosphere is the flow of radiation within the Sun. Since this energy originates in the center and flows outward, its intensity gradually diminishes at higher and higher levels inside the Sun. It is this difference between the intensity of radiation at different levels that would be available as free energy. The difference between the intensity at two levels one hundred meters apart deep within the Sun is greater than the flow of sunlight at Earth's surface. Also, the average energy of packets of solar energy is about the same as the energy of each charge within the plasma. This would make it convenient for plasma life to use radiant energy in its vital processes.

A real problem for plasma life is the absence of a solid surface in the Sun. Plasmobes would be vulnerable to being transported from regions where conditions were clement to other locations, such as the center, which are hazardous for them. We have already encountered this problem in connection with life in the clouds of Jupiter and Venus.

There are, however, several ways in principle of dealing with this problem inside the Sun. An object can avoid falling due to gravity by having the same average density as the surrounding material, thus achieving buoyancy. Having the same average density does not mean being identical to the surroundings, as can be seen from the case of a man floating in a swimming pool. For example, an object could have a more rarefied center and a denser outside than the surroundings and still have the same average density.

A second possibility is to utilize other forces to keep an object suspended. There are flows of material outward from the center of the Sun, somewhat like updrafts on Earth, and these can balance the gravitational pull on a dense object and keep it suspended like a glider. Alternatively, the strong magnetic forces acting within the Sun may be able to suspend some objects, particularly if they are electrically charged.

Finally, it should be recognized that some forms of life may not mind being tossed up and down inside the Sun because they may function equally well within very wide limits of environmental conditions. In fact, there may be advantages to the absence of a surface, in that this would allow free motion between widely different environments, making it possible for an organism to invent its own disequilibrium by moving from one condition to another.

How might plasma life develop if it gets started? Could the individual units, the plasmobes, speciate and become complex as has happened with Earthlife? This depends on what aspects of the system of charges and magnetic forces become ordered. The most direct way for life to become more complex is to increase its number of ordered units. In some sense, the reason that human beings are more complex than bacteria is that the DNA in a human cell contains a thousand times more bases than the DNA in a bacterium. Analogously, a larger collection of charge moving in some orderly fashion would be able to display more complex behavior than a small collection of charge. For example, the larger collection might be able to produce a greater variety of magnetic forces to influence its environment than the smaller collection, just as human DNA can produce more types of protein enzymes than bacterial DNA. Some of these magnetic forces might be arranged to repel other collections of charge, thus "defending" the organisms that produce them. Other magnetic forces might be able to attract passing charges to "feed" the present organism. These magnetic forces could be used to manipulate the environment and increase the supply of free energy by one of the methods discussed above.

The size of a plasmobe would be influenced by several factors, including the maximum amount of charge that can be effectively confined by magnetic forces, and the minimum amount of charge that can function as an ordered collection. Both these factors are difficult to estimate accurately. In a plasma like a star interior, it is possible for a magnetic force to act over very large distances so that even in a very large object, the different parts can exert an influence over each other. This suggests the possibility of plasmobes that are astronomical in size. However, it may be that a large plasmobe will tend to divide into many smaller ones because the magnetic forces are more intense over small regions than over large ones. The minimum size for a plasmobe, in terms of the number of particles needed to exert some self-stabilizing force, could be quite small, no more than a few thousand ions. But larger numbers might be needed to ensure that the configuration is stable against a wide variety of outside disturbances. The time scale for the life processes of plasmobes should be much shorter than for Earthlife of similar size because the physical processes in the plasmobes involve the motion of high-speed particles. Even though the particles do not move freely but are influenced by mutual collisions and magnetic forces, the time interval required for parts of a plasmobe to affect each other should be much less than for familiar organisms. Furthermore, the greater amounts of free energy available would have the same effect of speeding up the plasmobes' "metabolism," unless it contained many more particles than do Earth organisms. Consequently, plasma life could either develop in the direction of much more rapid evolution and metabolism than Earthlife, or in the direction of organisms that are much more massive than those on Earth. Perhaps both these directions have been chosen.

Plasmaland itself could include a large part of the interior of the Sun, but not the deep interior. In that area the forces associated with radiation flows and high temperatures are probably much greater than the magnetic forces. The presence of such large disruptive forces would make the region hazardous for plasma life. While one tends to think of the inside of a star as homogeneous, conditions in the deep interior actually differ more from those in the outer regions than those at the center of the Earth do from those at its surface. Life that is suitable for some regions of a star might find other regions as inhospitable as humans would find the center of the Earth. However, on Earth, evolution has produced advanced life forms that were able to colonize some areas which were previously uninhabitable. It is possible that evolutionary developments within the Sun have worked similarly to produce beings capable of penetrating the central region. If so, these beings might be able to tap an additional source of energy—nuclear fusion.

It is only in the deep interior of the Sun and of other stars that the temperatures are high enough for fusion reactions to occur. The rate of such reactions can be increased by raising the temperature of the appropriate region, by concentrating the matter there through magnetic forces, or by increasing the concentration of the specific nuclei that serve as fuel. The type of plasma life we have been discussing should be capable of controlling all these factors. The energy from the fusion process would appear in the form of subatomic particles of the same type that are in the Sun already. This would pose a technical problem for any life there. The energy released in an individual fusion reaction would be much greater than the average energy of the particles inside the Sun, so it might be difficult to couple this energy to the internal order of plasma life. The problem would be analogous to the difficulty of Earthlife using an intense flow of X-rays as a source of free energy. If such problems could be overcome, then nuclear fusion could be a useful supplement to the radiant-energy flow available to sustain plasma life. Fusion energy might serve as the "food" for specialized organisms, which might then become part of an integrated biosphere, as has happened for some energy sources on Earth.

If we disregard the possibility of colonization of the center, and Plasmaland is restricted to the outer regions of the Sun, it would still constitute a biosphere immensely larger than that of Earth or any planet. The region involved is three-dimensional rather than two-dimensional, and contains a thousand times as much material as the whole solid Earth, let alone the thin surface layers that comprise the Earth's biosphere. There are of course many other stars like the Sun in the Universe. If the matter in many of them has in fact developed into life, then the scope of such life will be immensely greater than anything that can be achieved by chemical life confined to planet surfaces.

Despite this potential significance of plasma life in the Universe, it will be quite difficult for us even to learn of its existence, let alone the details of family life among the plasmobes. In Chapter 14 we will discuss possible strategies to tackle this problem in the future. Right now we want to shift our location dramatically from a very hot to a very cold spot.

LIFE IN THE COLD AND DARK—
THE CRYO-BIOSPHERE

The sun that shines on the hypothetical planet Cryobus is far less robust than ours. It is one of the cool stars that are quite numerous in our galaxy and emits only a feeble reddish glow. Cryobus itself is somewhat like our own Neptune, but is several times further removed from its faint source of warmth than is Neptune from the Sun. As a result, an unimaginable chill pervades the surface of Cryobus, on which the temperature is only tens of degrees above absolute zero. Most substances, even the familiar gases of our air such as oxygen and nitrogen, exist as frosty solids. Only the lightest elements, hydrogen and helium, have resisted this fate. As these are the two most common elements in the Universe, Cryobus contains a giant sea of liquid hydrogen molecules (each containing two atoms of hydrogen) under an atmosphere of helium and hydrogen. A few impurities have dissolved in the frigid hydrogen sea, but no other chemical variety exists. In fact, no chemical reactions at all take place, since molecules have too little energy to break bonds during collisions. The chemical composition of the planet and its sea is fixed.

Despite this state of affairs, small bits of living matter float in the sea. Some absorb the weak solar radiation directly, while others feed on nutrients in the solution. From time to time they reproduce themselves. They are made of almost pure solid hydrogen, with some internal trapped helium to keep them afloat (solid hydrogen sinks in liquid hydrogen), and an occasional impurity is present, such as oxygen. With apologies to the readers of J. R. R. Tolkien, we will call these beings H-bits.

Clearly, the life-styles of these organisms, their metabolism, and reproduction are not expressed in terms of our familiar chemical reactions. Fortunately, there are other internal properties which can be influenced both by collisions and by the long wavelength radiation that reaches Cryobus. Atoms in molecules can revolve and rotate in various ways, similar to dancers in a discotheque (Fig. 31). Like the dancers, the molecules will have more or less energy, depending on the amount of rotation. Cryobus is so cold that collisions between the cold hydrogen molecules will usually not suffice to start even the slowest rotation. This task can be performed, however, by the weak infrared and microwave radiation that reaches Cryobus from its sun, just as the ultraviolet radiation absorbed by molecules on Earth serves to break bonds not

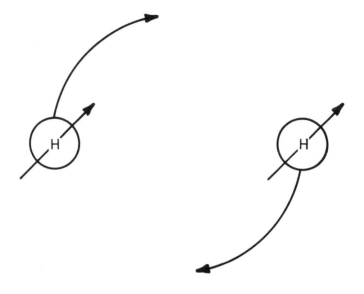

(*a*) An *o*-hydrogen molecule

(*b*) A *p*-hydrogen molecule

Figure 31. The two forms of hydrogen molecules are o-hydrogen and p-hydrogen. In the o form, pictured in (a), the two hydrogen atoms revolve about a common center, while at the same time their nuclei rotate about their own axes, in the same direction as one another. In the simplest version of the p form, pictured in (b), the atoms do not revolve, and the nuclei rotate in opposite directions.

normally broken in collisions. On Cryobus, hydrogen molecules will continually absorb energy, start to rotate, and then return to rest after emitting the energy or colliding with other molecules. The different forms of rotating molecules are chemically very similar in their properties. It is not chemistry that is involved in the life on Cryobus.

Rotating hydrogen molecules come in two types known as o-hydrogen and p-hydrogen. These molecules can be compared to pairs of dancers in which each partner spins around either in the same direction or in opposite directions, as well as revolving during the dance. (See Fig. 31.) Within each type there are other minor differences that are less important than the principal one which distinguishes the types. The o-hydrogen molecules rarely convert into p-hydrogen, though the various subclasses of each type can interconvert readily. In liquid hydrogen the less stable o-form will very gradually, over many hours, convert to the p-form, though the process can be speeded if certain substances (atoms or molecules that produce magnetic forces) are present. Unconverted o-hydrogen molecules comprise a reservoir of energy which is quite large (one hundred seventy calories per gram, about as much as cabbage) compared with the heat energies of the molecules on Cryobus.

Let us imagine bits of solid hydrogen floating in the sea of liquid hydrogen. Each hydrogen molecule in the solid lattice may be either in o-form or p-form. At equilibrium, almost all the hydrogens would be in the p-form, but an inflow of energy of the right type in the presence of other molecules as catalysts can convert many of the molecules to the o-form. Since the rate of conversion of o-hydrogen back to p-hydrogen is slow and is affected by the presence of impurities, there would be many opportunities for locally ordered regions of solid to develop which would be rich in one or the other type. This process would be analogous to the chemical specificity that developed among organic compounds in the early stages of Earthlife. While in Earthlife the order is described by the relative amounts of various or-

ganic molecules in the biosphere, and by the precise arrangement of these molecules in large polymers, on Cryobus the
order is described instead by the precise arrangement of
o-hydrogen and p-hydrogen in three-dimensional solid lattices. If such lattices occur repeatedly on Cryobus, with
o-hydrogens always in certain positions and p-hydrogens always in other positions, the situation can be as orderly as
that on Earth resulting from the repeated occurrence of
nucleic acid strands with a specific arrangement of bases.
The immobility of the atoms in solid hydrogen does not
prevent order from developing through absorption of radiation, and may in fact help preserve order by suppressing
collisions that could interconvert the two forms. Situations
which are a disadvantage for one mechanism in ordering
matter may be a benefit for a different mechanism.

A collection of ordered hydrogen molecules and impurities could form a primitive biosphere at low temperatures,
the cryo-biosphere. Such biospheres may be quite common
in the Universe, since cool stars are much more numerous
than any other kind, and most planets of these stars may be
as cold as Cryobus. A similar locale would exist on a planet
orbiting a white-dwarf star at the distance that Jupiter is
from our Sun, or on a planet in an immense orbit (a light
year in radius) around a cool supergiant. In order for life
to develop in the cryo-biosphere, it would be necessary for
specific properties of the molecules involved to act to increase the initial local concentrations of order. One suitable
property is the ability of o-hydrogen molecules to convert
to p-hydrogen molecules through the magnetic forces they
exert on each other. It is also possible for a rapidly rotating
p-hydrogen to convert to the o-form upon collision with
another p-hydrogen. Thus, molecules of the two types can
catalyze their own formation and destruction.

An H-bit, a more evolved bit of solid hydrogen, would
contain relatively stable arrangements of o- and p-hydrogen.
Thus, the outside molecules in the solid might all be stable
p-hydrogen with little rotation, which would not be likely

to convert to the o-form by collision. Such p-hydrogens could be formed by allowing solid o-hydrogens on the surface to collide with o-hydrogens in the liquid, with both converting. An H-bit could obtain o-hydrogens by allowing rotating p-hydrogens to collide with its magnetic-impurity molecules (such as oxygen). These processes are analogous to the enzymatic process of Earthlife.

As H-bits have a simpler "metabolism" than Earthlife, involving only a few types of molecules in a small number of configurations, they may be able to function with a smaller amount of order than living things on Earth. They may be no more complex than our viruses, or even smaller. A requirement for a specific shape and a minimum size comes from the need to shield those parts containing the less stable o-hydrogen from the environment (except when they are needed to catalyze a process on the surface). These internal o-hydrogens must also be kept apart from one another lest they catalyze their own destruction. The o-hydrogens would best be located midway between the center and surface of the H-bit, and not densely clustered (Fig. 32). This implies a

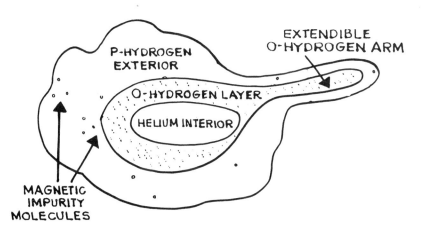

Figure 32. What an H-bit might look like. The hydrogen is in solid form, while the helium in the center is a liquid or gas. The extensible arm is used to catalyze the formation of *p*-hydrogen from *o*-hydrogen in the environment. With the addition of some magnetic impurity molecules, it can also catalyze the opposite process.

minimum size of at least several times the spacing of molecules in solid hydrogen. The minimal H-bit would contain about a thousand molecules and be about the size of a large protein (10^{-8} meters, or about half our height when we were on the -8 level of COSMEL). The mass of this H-bit would be much less than that of the protein, since hydrogen is the lightest element.

Mechanisms for growth and replication would be useful to help us further in recognizing the H-bits as a type of life. One such process would involve the crystallization of additional hydrogen molecules onto the surface of the H-bit. If the H-bit were arranged so that some parts of it were magnetically active and some inactive, it could control the pattern in the newly formed solid, arranging directly for the production of a duplicate of itself. An alternative process could take place when an H-bit encountered an existing but disordered piece of solid hydrogen. By positioning parts of itself near the solid, it could induce order into it. The new H-bit would not necessarily be a duplicate of the original one, but might itself have catalytic activity. These active H-bits would continue to induce order in random H-bits, while the inactive ones would remain inert until they were converted to active form. In this way the catalytically active H-bits would come to dominate the cryo-biosphere.

Of course, all these processes would require free energy. Those H-bits that lie on the planetary surface or float in the ocean could use solar radiation directly. Others, in the depths of the ocean, might survive on a surplus of o-hydrogen molecules formed by the radiation and stored in the cryo-biosphere. The amounts of energy necessary to sustain H-bit life are probably significantly less than for our own. The processes involved in H-bit metabolism involve intrinsically smaller energy changes. Less energy is also needed to maintain order at the low temperatures present on Cryobus. Furthermore, the properties of liquid hydrogen would make it substantially easier for a bacterium-sized object to move about in it than in water. In all, despite the low-energy content of the radiation from the sun of Cryobus, free energy should not be a serious problem for H-bit life.

What course would evolution take on Cryobus? A multi-cellular structure, analogous to that on Earth, would provide similar advantages to those we enjoy. Differentiation into organs that specialize in radiation absorption, o-hydrogen capture, hydrogen-molecule conversion, acquisition of helium for flotation, and so forth would be possible. Some of the gross features of Earthlife might thus be duplicated by a cold life form based on a completely different physical basis. Other facets, of course, would be very different. As the structures involved in life on Cryobus will be based on differences in the rotation of atomic hydrogen, small H-bits and inanimate bits of solid hydrogen will look alike under visible light. An intelligent H-bit would also have a problem in deciding that an Earth bacterium was alive, given the absence of any order in the rotation of the atoms in the bacterium's molecules, just as we would have trouble in detecting life in the H-bit. In either case only by an analysis of behavior, or of the exact arrangement of the molecules within the solid, could one type of life recognize the existence of life in the other type.

RADIANT LIFE IN (ALMOST) EMPTY SPACE

In our last stop, we will revisit the interstellar clouds of gas and dust encountered in our tour of the galaxy. For purposes of contrast we will consider two such clouds. One, which we will name Chaotica, is a typical cloud of average density, remote from the center of the galaxy. Its properties are similar to the clouds described in Chapter 11. The other cloud, Radia, is denser and closer to the galactic center. Most notably, it has been transformed by the presence of a life form—radiant life.

Even in a relatively dense interstellar cloud such as Radia, the atoms and molecules comprising it are quite distant from one another, so that collisions between these particles are very rare. The density is as if a large grain of sand were in front of us and its nearest neighbor were as far away as Moon. While the atoms in the cloud do not interact easily through collisions, they can influence one another by exchanging radiation. Consequently, the process of life that takes place in Radia is profoundly different from anything in our experience with Earthlife, or even contemplated in most fantasies. It is based on the properties of ordered radiation, using matter as a necessary intermediary.

We can best appreciate the properties of radiant life if we first consider how radiation behaves in the lifeless cloud Chaotica. Radiation of various types is emitted by nearby stars and absorbed by the atoms and molecules in the cloud. Different atoms and molecules emit and absorb specific wavelengths. As we have seen, interstellar clouds contain a variety of atoms and molecules, each of which may respond to different wavelengths.

One effect of radiation on atoms is to cause them to lose electrons and become charged ions. It is another process, however, that is most relevant to radiant life. Atoms and molecules can exist in various energy levels. The presence of radiation will influence the number of atoms in each of the levels. This is a more specific effect than ionization because there is a close relationship between the type of atom or molecule and the type of radiation that can influence the population of all its energy levels. The radiation that is emitted and can be absorbed by an atom or molecule is as characteristic of that object as the types of chemical reactions it can undergo.

In Chaotica the atoms and molecules present are distributed randomly among their various possible energy levels in a pattern close to what we have called equilibrium. This random distribution of atoms will emit and absorb radiation in a wide variety of wavelengths, so that little order is present either in the atoms or in the radiation.

A very different situation prevails in Radia, which contains a large organized region, the "radiobiosphere." The location of Radia near the center of the galaxy is advantageous for life, as it ensures that a plentiful supply of stellar radiation is available as "food" for the radiobiosphere. While other energy sources could conceivably be used by radiant life, this one is obvious and convenient. Because of its density, Radia is quite effective in trapping the photons that hit it. If we assume that hydrogen atoms are the main absorbing species, then for any cloud we can estimate the minimum size it would need to ensure that every photon is absorbed. The answer also depends on the wavelength of the incident photons. At the ordinary interstellar density of hydrogen atoms, a region about one light year in diameter would be needed to absorb each photon of ultraviolet light. On the other hand, in a relatively dense cloud with almost a hundred million hydrogen atoms per cubic centimeter, a region with the radius of Jupiter, about 10^8 meters, would suffice. It is interesting that in a dense cloud, a smaller total number of atoms is needed than in a diffuse one. The cloud with interstellar densities would contain about as many atoms as the Earth, while the denser cloud, such as Radia, would contain only about one hundred million tons of material, about the total weight of the whole human race. It is possible that a minor component of the cloud, rather than hydrogen, could be the principal absorber of external radiation for the radiobiosphere. In that event the size of the cloud would necessarily increase to compensate for the diminished concentration of the absorber.

Within the radiobiosphere of Radia, a much more ordered situation exists than inside Chaotica. Fewer types of atoms and molecules are present, and these absorb and emit radiation more selectively. Furthermore, the atoms and molecules are distributed in a more concentrated way among their many possible energy levels than in Chaotica. This results in a pattern of emitted radiation that is also concentrated in a few of its many possible wavelengths. It now resembles the radiation produced by a laser, rather than the broad spectrum present in sunlight (Fig. 15). These laser-like emissions are in turn reabsorbed by other objects within the radiobiosphere. A complex ordered pattern of radiation within Radia results. Furthermore, because the wavelength of this radiation also depends on how the objects emitting it are moving through space, order has developed in the motion of the atoms and molecules of the radiobiosphere.

Precisely which wavelengths are used by the life in Radia depends on the matter present and on the external energy source available to it. The photons emitted by radiating atoms can be much more energetic than those emitted by molecules. It is most convenient for the operation of the radiobiosphere that the photons of its external energy source contain more energy than those used in the transactions of radiant life, because it is usually easier to divide a high-energy photon into several lower-energy ones than to do the opposite. So a radiobiosphere such as that in Radia, based primarily on atoms, would need a source of external energy which comes in large packets, such as ultraviolet light or X-rays. In another cloud where molecules are the dominant form of matter, external sources such as visible, infrared, or microwave radiation could be used.

Although it is possible that a radiobiosphere could exist as a single integrated unit, within Radia a subdivision into smaller, independent units has taken place. Each unit is self-contained, in that it reabsorbs most of the ordered radiation emitted within it. Eventually, of course, a radiobiosphere and each of its units will emit as much energy as it absorbs, but the emitted radiation will be in a less available form, allowing order to increase within the system. If we assume that the radiation involved is visible light, we can calculate the minimum size needed to constitute such a self-contained unit. In a dense cloud like Radia, it measures only one centimeter and contains only one million atoms. On the other hand, if radiant life were to develop in Chaotica with its much lower density, the minimum size for self-containment would be about a thousand kilometers, a tenth as big as Earth.

Such subdivisions are the individual beings of radiant life. They are small enough that radiation emitted anywhere within them can reach any other point in a short time, less than one second for the largest of them. We will call them "radiobes." We can imagine a life cycle for a radiobe. It is "born" as a collection of atoms or molecules in a specific pattern of excitation. These excited atoms, under the influence of extra radiation from outside, gradually emit their own radiation in an orderly pattern. This radiation, interacting with the same group of atoms, reexcites some of them into yet another organized pattern, and this cycle of exchange of order between radiation and atoms continues, perhaps at some point pausing to emit a beam of radiation which can begin the ordering process in a nearby lifeless collection of atoms. If this process appears too physical to be considered a type of life, readers are invited to consult a similar description of the information transfer between nucleic acid and protein in Earthlife and decide whether that process is any less physical.

It is interesting to estimate how much radiation is contained in a radiobiosphere and individual radiobes. One measure of this is the number of photons in a cubic centimeter of biosphere. This depends somewhat on the details of how the photons are emitted and absorbed; but in highly evolved radiobes the photon density should approach 10^8 per cubic centimeter, similar to the density of atoms in a thick cloud. This is about the same as the photon density in the sunlight near Earth. However, that light is diffused over a wide spectrum of wavelengths, while in a radiobe the photons would be concentrated at a few specific wavelengths, which accounts for the much greater amount of order present.

A feature of the interaction between radiation and matter which aids in establishing order in the radiobiosphere is stimulated emission. Suppose an atom is in an excited condition from which it can emit a specific wavelength of radiation. If there is already some radiation of the same wavelength present nearby, the atom is more likely to emit its radiation than if it were isolated. It is somewhat like the human tendency to start yawning when those around us yawn. We have mentioned this effect in connection with lasers, as a way in which the diffuse energy input to the laser becomes concentrated into the narrow band of wavelengths the laser emits. This is a primitive form of the ordering process we are imagining for the radiobiosphere. It is also possible for the opposite effect to occur. Atoms whose arrangement is especially orderly can decline to emit as much radiation as the individual atoms would, thus prolonging the existence of the atoms in an excited condition.

How would a radiobiosphere evolve? It might start from an unordered distribution of atoms in space, together with the radiation they emit and that impinging on them from elsewhere. If the atoms and molecules present, the energy supply, and available time were all suitable, an ordered state could arise and influence the further development of order. The general principles involved would be those that influence the evolution of all life forms.

As the radiobiosphere continued its evolution, it could increase the number of ways in which it influenced its environment, just as Earthlife has. Radiation patterns could possibly influence not only the excitation of atoms and molecules but even what molecules exist. Interfering molecules, ones that absorb radiation essential to the operation of a radiobe, could be destroyed by using the proper radiation. The gravitational contraction of interstellar clouds, which ultimately leads to star formation, could threaten the existence of the radiobiosphere. Part of the cloud could be stabilized against such collapse by the outward pressure of the radiation emitted by the radiobiosphere. Such properties might be useful in helping us detect the existence of radiant life.

Some readers may have difficulty in considering that a collection of isolated atoms and molecules in space, which absorbs and emits complex patterns of radiation, is a life form at all. We have indicated that it has the same constraints and meets the same requirements as other life forms we have discussed. It requires an energy flow from outside. Given this, it can increase in complexity and store its order. It may even form individual organisms. The difficulty in recognizing it as life would be due to our own psychological limitations, rather than to the principles that are involved.

The living beings of Earthlife function primarily through the interactions of matter with other matter. Some processes of Earthlife, such as photosynthesis, involve the influence of radiation on matter, but this influence is one-sided, in that the living matter does not affect the radiation except to absorb some of it. Matter is the major component of the life we know and of any life forms that have usually been imagined. It would be quite fitting however that electromagnetic radiation is the basis of a form of life. It is in many respects as significant a component of the Universe as what we know as matter. It is more widely distributed than matter, and in some circumstances in the center of hot stars, radiation can even be as densely packed as matter is on Earth.

In radiant life, radiation acts on matter in a way that causes it to emit related radiation. In other words, matter is a tool used by radiation to produce order in itself, in the same sense that proteins are a tool used by nucleic acid in Earthlife to replicate itself. It is probably more correct in either case to consider both components as an essential part of the life process, but we have used the term "radiant life" to emphasize the novelty of ordered radiation within the life form.

It appears less likely that radiant life, completely free of matter, could exist. Radiation, unlike matter, is for the most part indifferent to itself. While two chunks of matter that are near one another can affect each other's behavior through collisions, two independent flows of radiation will usually pass through each other without having any significant effects. The role of the lonely atoms and molecules in Radia appears vital.

On the other hand, when there are substantial amounts of matter in a locality, the interactions of the matter with itself would obscure the effects of the radiation. Interstellar space, where radiation is plentiful and matter is sparse, is the ideal setting for radiant life.

In our comparison of Radia and Chaotica, we have suggested two factors, energy supply and density of matter, that might lead to the evolution of a radiobiosphere in one locality and not in the other. Other factors such as time may also play a vital role in determining whether such life develops.

We do not know how long it would take to evolve radiant life in a suitable region of interstellar space, or whether any radiobiospheres exist at present. Radiation processes are usually thought of as extremely rapid, but that is because we generally observe them in a small region of space. Even light takes years to travel over the astronomical regions that comprise early radiobiospheres. Other slow physical processes, which we have not considered, may hinder their development.

For these reasons we cannot be certain that radiant life exists in the Universe. The same is true for the other forms of physical life that we have discussed. Such definite predictions would go far beyond the current abilities of theoretical science. The examples we have given were meant to demonstrate that life forms based on physical, rather than chemical, processes are possible. If we are correct in thinking this, then it is likely that such life does exist and makes its home in such common environments of the Universe as stars and dust clouds. It is less important whether such life is precisely of the types we have discussed than whether some does exist. We can only learn this by further observation, exploration, and deduction.

Chapter 13

The Creatures of Elsewhere

In this book we have concentrated on the underlying principles and mechanisms by which life in other places than Earth might function. We have also considered the types of biosphere that could develop in these places. The first topic is equivalent to the subject matter that molecular biology deals with in studies of Earthlife, while the second is related to ecology. Most past speculations and fictional accounts of extraterrestrial life, on the other hand, have emphasized a question that falls between these two; that is, one that lies within the scope of zoology and botany: What creatures do other worlds have to offer compared with the ants and elephants, humans and Sequoias of Earth?

One of the remarkable things about Earthlife is the tremendous variety of creatures that have evolved, all sharing the same underlying mode of chemical storage of order. On Earth superficial differences in appearance or behavior mask this underlying similarity. When the basic living process is very different, we might expect great variations in appearance and gross behavior from the creatures of Earthlife. But eerie parallels between Earthlife and other forms may exist in spite of great differences in their bases of life. An analogy to this exists on Earth also. While mammals, birds, and reptiles differ significantly in their physiology, each order of creatures has evolved flying types whose behavior is much more similar than the differences in structure might suggest. There may be fewer things that life can do than there are ways of doing them.

We will consider the question of highly evolved extraterrestrial life, using the principles we have described and the other known laws of science as a guide rather than our unrestrained imaginations. This is a much harder question than the ones dealt with so far. It is very unlikely that the features of the many species that have lived on Earth could be predicted from a knowledge of the types of chemicals that occur within our cells. Such predictions are even less likely to succeed in the case of a speculative form of life that has never been observed directly. What we can do is examine Earthlife and consider which of its large-scale features are essential and which are merely local solutions to a general problem. For the latter we suggest alternatives that may occur elsewhere in the Universe.

THE QUESTION OF ORGANIZATION

Earthlife is divided into subunits which themselves are divided into smaller entities. This compartmentalization continues over a number of levels. We are familiar with the various structures involved: species, individual organisms, organs, tissues, cells, and, finally, cellular organelles such as mitochondria. This subdivision, for administrative purposes, is characteristic not only of our biosphere but also of many other complex systems on Earth: governments, economic systems, universities, and so on. It may be a vital need of any system, once it reaches a critical level of complexity, to subdivide for purposes of efficiency. It is hard for a single center to control larger and larger outlying regions effectively. But while we know from experience that subdivision is probably inevitable, the details involved vary widely from system to system. A free-enterprise economic system, for example, differs considerably in structure from a socialist one. Similarly, other biospheres may vary greatly in their organization from the one on Earth.

Our own biosphere appears closest to a free-enterprise system in which fierce competition takes place. Some species fail and go out of business, while others evolve and take their place. We cannot predict the exact list of which companies will be delivering milk fifty years from now or how this will be done, yet the milk will be distributed. The crucial jobs of the biosphere, such as photosynthesis and nitrogen fixation, get done as well, though the species that perform them may change with time.

A socialist-type biosphere can also be imagined, in which the activities of every living unit are closely interlocked and under some unified control. In the extreme case, the units may be as interconnected as they are in a single living thing on Earth, so that the entire biosphere may best be considered as a unified organism or cell. Such a macroorganism could be of planetary size if it evolved in the right environment. If it were a plasmobe or a radiobe, as discussed in Chapter 12, it could be even larger and cover much of the interior of a star or galactic cloud. Whether such large integrated macroorganisms occur will depend on their ability to extend control over their whole size. This will vary with the underlying structure of the biosphere. For Earthlife the trend toward multicellular organisms may have resulted in part from the inability of a nucleic acid strand to send its chemical messages over large distances. The same is true on higher levels of organization, and acts to limit the size of organisms because of the difficulty in controlling behavior by the central nervous system. Even the size of nations has in the past been limited by ineffective communications over great distances. The latter problem has partly been solved by the use of radio. Similarly, in a radiobe or plasmobe the distances over which messages can be sent may be much larger, and the largest effective size of a cell or organism would grow accordingly.

Of course, such an integrated biosphere runs the risks of putting all its eggs in one basket. The death of the individual would equal the extinction of life in that region. Risks could be minimized, however, by duplication or multiplication of some internal organs, just as we have two eyes and kidneys in our own bodies. A planetary organism would not replicate within its own realm, though it might attempt to release seeds to form colonies on other worlds. In Chapter 6 we discussed how replication and the birth and death of individual organisms are an effective strategy for increasing order and responding to environmental change, but not the only one. A macroorganism might live in a rather constant environment, where competition and the extinction of individuals and species would be wasteful. Repair processes and partial reconstruction would suffice to maintain order and introduce innovations.

Even in those cases where other biospheres are subdivided into competing units, dramatic differences from our own type of organization may occur. The cells of higher organisms on Earth all derive from an initial single fertilized germ cell. For the most part, each of these cells contains a full copy of the genetic message of that original cell. The DNA of the early cell had all the information needed to make livers, skin, nerves, muscles, and other tissues and organs. Each specialized descendant cell executes the portion of the message relevant to its function and disregards the rest. If a single liver and skin cell were taken from a large number of individuals and all the cells were mixed together, it would be possible in principle to sort them into pairs corresponding to each individual just by examining their DNA sequences.

Another situation could prevail elsewhere. Organs could exist which contain only the information for their own function and which can replicate independently. They could not live for long on their own but would have to function in combinations. A similar state of affairs exists with lichens on Earth, but it is not very common or obligatory. In such an alternate biosphere, organisms of fixed form would not exist continuously, but could be assembled or disassembled from organs as the occasion required. If an organism wished to cross water, for example, it would acquire appendages for paddling and floating, which it could later shed. A creature might retain as many appendages of different types as it could induce to join it and keep nourished and content within the union. The hybrid creatures which on Earth are a feature only of mythology would be the rule of the day.

The above combinations would presumably be unions of equals. In another arrangement, we can envision a core of an individual which had a full copy of the genetic material, while individual organs had only their relevant sections. For adaptability the entire genetic messages of the organs might have only a finite lifetime and be replaced periodically by new orders from the core, somewhat as a snake sheds its skin each year. When replication or other forms of genetic diversification occurred, only the core need be altered. The organs could remain intact and await reprogramming, rather than be discarded wastefully, at death, as is the case on Earth. Possibly, cellular structures could be built on a modular basis, with cell walls and organelles that could be disassembled and moved about to aid the process of reprogramming.

Another aspect of organization is the connectivity of an organism. It is possible that an organism could be made of parts that are not joined, but still function as an individual unit. A single part, viewed in isolation, might be mistaken for a separate organism, but its behavior would be influenced by, and integrated with, the other parts of the larger unit. A model for such a system might be an ant colony or a beehive on Earth. The extremely integrated behavior of individual members has led some observers to consider the entire colony or hive as a type of organism.

Creatures in other locations might have evolved naturally in this direction. The plasmobes which we suggested might inhabit the interiors of stars are natural candidates, since magnetic forces would be the catalysts for their activities. A "plasmobeast" might consist of a central pattern of moving charges held together by its own magnetic forces, and of detached sections of charges held together and moving according to forces generated by the central pattern. The outlying sections could act as gatherers of energy and of certain ions that are necessary to the plasmobeast's metabolism. The number of such extensions and the size of the whole plasmobeast would depend on the environmental conditions in which it lives, since magnetic forces weaken when they extend in all directions over great distances.

TIME AND THE BIOSPHERE

Our biosphere is perhaps three and a half or four billion years old. It could even have an important birthday coming up, but if so no one would know when to celebrate it. We humans, who care about such things, unfortunately share the fate of other organisms in Earthlife. We are perishable, with a life span much shorter than that of the biosphere.

The relevant time spans for creatures on Earth vary from minutes to years. The time that any living thing takes to perform its functions depends on the physical processes through which it operates. The faster these processes are, the faster the rate of life of the creature. Among living things on Earth, the rate of life tends to correlate roughly with the size of the organism. The larger it is, the slower its life processes. A bacterial generation can occur every twenty minutes, while for humans it may take half a million times as long.

In another biosphere, a creature the size of a bacterium might take much more or much less than twenty minutes to reproduce or carry out any other living function. For chemical life the time would vary with the temperature and with the reaction rates for the key processes. Life with a different chemical basis might function much more quickly or slowly than Earthlife, as chemical reaction rates vary tremendously from one type of reaction to another. Physical processes that might be the basis of life forms vary even more in their rates. We can therefore expect a wide range in the rates at which living things inhabiting other biospheres function.

It might be hard for us to recognize the existence of individuals whose life spans lasted either fractions of a second or millennia. Other difficulties in recognition would occur in cases in which individual lives consisted of alternating dormant and very active stages. Lichens in the Antarctic and in Alpine locations show intense bursts of activity on those infrequent occasions when conditions are favorable for life, but exhibit minimal activity at other times. As we mentioned earlier, tardigrades and spores can exist in a desiccated state for long periods of time, yet return to active life when water is supplied. Biospheres may have developed in locations such as planets with irregular orbits, or in multiple star systems, where most of the time conditions are unsuitable for life processes. To compensate for this, the basic life processes might take place very rapidly when things were favorable. This would allow much evolution to have occurred. A casual examination of the biosphere at a randomly chosen moment would detect no activity, however. All living things would be encapsulated or otherwise preserved to protect their existence. In such cases it might require much ingenuity to recognize that we were inspecting a site for life.

STRATEGIES FOR ENERGY

Earth's biosphere has been lucky. Our principal source of energy, the Sun, has been reliable, and sunlight is available in abundance. The primary receptors of this energy, the plants, have only to expose passively a suitable surface toward the Sun in order to receive energy. The surface of the Earth has provided more than ample space for this, and the biosphere has been able to pick and choose among sites for plant life. The amount of sunlight may influence the size of plants, while other factors such as the availability of water and nutrients have been the ones that limited growth. Where these are insufficient, plant life has not flourished, even though the energy for photosynthesis has been available.

The type of energy used, its availability, and the manner in which the primary receptors are deployed to receive it will be important features of any biosphere. We will use the word "plants" to describe these receptors, whatever the energy source and whether they are mobile or not. Thus, on a planet such as Venus where thunderstorms are an important energy source, plants may come equipped with organs that act as lightning rods. We have discussed worlds like Jupiter where temperature differentials may be used for energy, and we used the name thermophage for the creatures making use of these differences. In a more tranquil environment, thermophages may adopt a passive approach resembling that of Earth plants. One end of the thermophage may be fixed in the source of heat, the other in the sink. A connecting part, perhaps several kilometers long, would serve to transfer heat from source to sink. Such extended plants may also exist in biospheres using conventional (to us) solar energy. If the atmosphere of a plant were thick but tranquil, evolutionary developments might lead to the extension of plant receptors to the very top of the atmosphere at the edge of outer space. Nutrients could then be supplied through "roots," which would trail far below to the surface of the planet.

This leads us to the subject that has received the most attention in the literature on extraterrestrial life—the actual shapes and behavior of living things. What can we say about this subject from our own pespective?

ANIMALS OF OTHER WORLDS

One ambitious effort to imagine the forms of extraterrestrials is displayed in the National Air and Space Museum in Washington, D.C. The exhibit describes nine planets with various temperatures, gravity, and water content. A possible sea, land, and air inhabitant has been conceived for each planet so that twenty-seven animals in all are displayed. The beasts differ in size, position of organs, number of limbs, and other visible features from those we know on Earth. The resemblances to each other and to Earth animals are more striking, however. Arms, legs, heads, teeth, wings, and eyes are prominent. All of them require oxygen and water and presumably share our biochemistry. In fact, similar beasts may well occur somewhere in the Universe, but the collection is hardly a cross section of the possible varieties of extraterrestrial life, no more than a dog show represents the range of Earthlife.

If beasts of other worlds are to be shown in shapes similar to those on Earth, then it is only natural to carry the process further and display alien intelligent life in human or humanoid form. We have frequently seen creatures of this type in science fiction literature and films. The benevolent beings of *Close Encounters of the Third Kind;* Chewbacca the Wookie, the Jawas, and the Sand People from *Star Wars;* and many, many others come immediately to mind.

If this effect were limited to films, it would simply reflect the problem of casting actors in unusual shapes and forms, and the cost of building special animated robots. However, such concepts have also entered scientific literature. We will quote from a book by space scientists Roger A. MacGowan and Frederick I. Ordway III:

> *Travel over rough and variable land terrain probably would dictate the development of four legs for the majority of large land animals on other planets. Later evolutionary developments on planets having dense forests (any planet having prolific life could be expected to have plant life similar in gross characteristics to that of the Earth) would almost inevitably lead to the conversion of forelegs to arms in some of the moderate sized animal species. . . . It may be concluded that the majority of intelligent extrasolar land animals will be of the two legged and two armed variety.*

The difficulty with such assumptions is that they do not consider alternative possibilities and suggest means to distinguish among them. They represent yet another extension of the predestinist thinking discussed in Chapter 5. In fact, the weight of evolution theory stresses the unlikelihood of the appearance of humanoid forms elsewhere in the Universe. The arguments have been made at length by the Nobel Laureate F. Jacob and by evolutionist George Gaylord Simpson. They point out that the exact course of evolution has been influenced both by random genetic events, such as mutation, and by local environmental factors, which influence life through the mechanism of natural selection. Slight changes early in evolution would accumulate and lead to greater divergences later on. It would be no more reasonable to expect that a man has been produced by independent biological processes on another planet than to anticipate that the exact shape and contour of New York Harbor would be re-created in detail by geological processes elsewhere.

Earlier, we discussed very strange possibilities for extraterrestrial life, such as plasmobes, H-bits, and lavobes. Still, we are animals, so it is natural to wonder what our equivalents might be like on other planets. By animals, we mean organisms which do not use the primary source of the biosphere directly, but use instead the "plants" that do so as stores of free energy. One way to accomplish this is to move around and eat whatever plants are in your path. A question that springs to mind is whether Earth animals and extraterrestrial animals could eat each other. If we did try to eat them, or vice versa, the most likely result would be indigestion. Their chemistry is probably quite different from ours, with perhaps a few small molecules shared by chance. Small changes within large molecules on Earth can be sufficient to alter foods we can digest, like starch, into ones we cannot, like wood. In extraterrestrial beings there will be many chemical differences. The effect of munching an extraterrestrial "carrot" could probably be approximated by sampling the contents of a chemistry lab at random. A few chemicals may make us sick, others may accidentally taste sweet or bitter, but many will have no effect at all.

An extraterrestrial made of exotic chemicals may still have some appendages like those we know on Earth. The form of the organs will often be dictated by the structure and physical laws involved in their function rather than by their building materials. Thus, stone, wood, and grass houses all have doors and roofs. Birds, bees, and bats all use wings to fly, though they have evolved by different paths. The term "convergent evolution" is used to describe this phenomenon on Earth, and we will undoubtedly encounter it on other worlds.

On the other hand, worlds that differ from ours in their physical environments may produce creatures which would appear strange to us even if their biochemistry were not very exotic. Thus, very large animals may occur on planets with weak gravities. Even on worlds with substantial gravity, large animals can live "floating" in liquids or gases, as do Earth whales. In such a case the maximum size of the animal may be limited by the distance over which communication is possible between its different parts. However, there are environments such as the surfaces of white dwarfs or neutron stars where the gravity is immensely greater than on Earth. In such environments life is likely to be very small indeed in height, although it can be large in other dimensions. Such creatures could be so thin as to be invisible from the side, even though they appear huge when seen from above.

IS IT TECHNOLOGY OR BIOLOGY?

One great attraction of extraterrestrial life is the possibility of novelty. In science fiction and fantasy, the author may give extraterrestrials any attributes that he or she fancies. We would rather stay within the limits of things that we think are possible. One continuing source of novelties for human beings has been our own technology. Often our discovery of a tool is followed by the realization that it is already employed in Earthlife. Birds use our planet's magnetic field for navigation, bats guide their flight by sonar, and electric eels were shocking their prey long before the days of Franklin and Volta. Not all the tricks of our technology are utilized by Earthlife. Yet those that have been missed here may well have been discovered elsewhere, the result of natural evolution rather than technology.

Wood and bone are used by Earthlife as common structural materials, for example, but not iron and plastics. It has required the ingenuity of mankind to discover the latter substances. It is conceivable, though, that organisms in another biosphere could extract metals from their ores metabolically, or that a hydrocarbon-based life could have made use of polyethylene. If an extraterrestrial being with an aluminum or plastic covering approaches us, it will not necessarily indicate that it has an advanced intelligence. It may instead be an indication of an alternative biochemistry.

Devices as well as materials may have evolved on other worlds. We have mentioned hydrogen-filled balloon beings as possible inhabitants of Jupiter. On Earth, communication by sound and chemicals is common. Extraterrestrial beasts may instead use microwaves or radio waves with home-grown antennae and receivers, especially if such radiation is a primary energy source on their planet.

We will end the list with these few examples, although a much longer one could be made. Our point is not that any of the above features must occur, but that they could naturally occur elsewhere, having been shown to be possible by technology here. This technology, of course, has by no means exhausted the list of devices to be invented, nor has science completed its effort to understand the laws of nature. It is quite possible that we will encounter principles and devices at work in other creatures not yet found here on Earth, though we cannot yet state what they may be. Much has been written about the appearance of intelligence as the pinnacle of evolution on Earth, and about the possibility of intelligence existing elsewhere in the Universe. It is even more tempting to ask: What *alternatives* to intelligence may have arisen as the prime achievement of evolution in other biospheres?

There may be many wonders awaiting us, if we care to make the search.

Chapter 14

A Universe Full of Life and How to Find It

The paths we have taken in our study of life have led us to the conclusion that life is a widespread feature of the Universe rather than something confined to the surface of one insignificant planet. We have found that many different properties of matter and energy can evolve into forms of life, not just certain chemical combinations of a few elements within a narrow range of temperatures. Even though the Universe as a whole may be growing less orderly, order can increase locally because radiation and matter move about, transferring order out of some places and into others. It seems to be a general aspect of how the Universe evolves that orderly arrangements of energy and matter develop in some regions, while other regions become less orderly. Processes as different as the formation of stars and the evolution of Earth's biosphere are examples of this tendency.

We see life as being an integral and essential part of the Universe, a preferred expression of possibilities inherent in natural laws rather than a rare accident never to be repeated. Such conclusions are reassuring to those of us who find life to be an interesting study in itself, as they immensely extend the range of life's possibilities. The conclusions are perhaps deflating to any who think of Earth and Earthlife as something unique and distinct from the rest of the Universe, or as the purpose for which the Universe was created. In this respect our conclusions follow the same tradition as many earlier scientific revelations, such as the Copernican discovery that the Earth revolves around the Sun.

WHO CARES ABOUT EXTRATERRESTRIAL LIFE?

The view that life is widespread in the Universe is theoretical and highly speculative, though it is based on the best scientific information presently available. But scientists cannot be permanently satisfied with a purely theoretical basis for important conclusions. If we really want to know the truth about extraterrestrial life, we must eventually seek it in some of the places where it may be found. The ultimate test of any scientific idea, no matter how reasonable it appears or how well founded in accepted principles, is whether it leads to suggestions for new observations or experiments that confirm it.

Not only scientists are interested in extraterrestrial life. People in general find that the forms of life present in any environment are one of its most interesting aspects. Many of the comments about Moon made by astronauts who have been there stress its deadness. We think that some of the dropoff of interest in Moon after the first expedition can be traced to the lack of any life there. If, on the other hand, some types of life are there to be found in various extraterrestrial locations, an upsurge of popular interest in exploring the locations and finding out what is there would certainly result.

The conclusions of this book also are relevant to the question of long-range goals as the human race advances toward the twenty-first century. If our civilization continues on its present course, we will be faced with many choices involving our intentions toward the rest of the Universe. Very different sets of goals may occur to us if we learn, on the one hand, that life is extremely rare and we are essentially alone, or on the other, that the Universe is alive with fascinating and diverse living beings. The question concerning extraterrestrial life is one which we will want answered before we come to more final decisions on our goals and the meaning of our existence.

All these are reasons why we should continue the search for extraterrestrial life. The question is, how can we best do it? At the present stage of human technology, it is obvious that looking for extraterrestrial life is not a project for individuals or societies that are impatient for results. Even within the solar system, the scale of distances is so vast compared with the speeds of space vehicles now available to us that the exploration of any of our neighbors other than Moon would require years or generations to accomplish. To this travel time must be added the years required for agreement on the desirability of any given exploration and on its design. It is clear that the search for life within the solar system will require generations, if not centuries, to carry out even in a limited way. Perhaps we may compare it to the exploration of the Western Hemisphere after Europeans visited it in the fifteenth century, which also took several centuries to complete. This example also indicates that the human will to carry out projects over long periods of time cannot lightly be dismissed.

The time scale may shrink in the future as various technological and social developments make extraterrestrial exploration faster and easier. From the perspective of one or two centuries in the future, our accomplishments in space exploration so far will probably seem about as primitive as the voyages of Columbus appear to us now. It is difficult to anticipate precisely how and when these technological developments will occur, but advances in space technology are one of the surest bets for the future.

We will divide our suggestions about how to look for extraterrestrial life into several parts. Both the short-term and the longer-term projects will require not only technological advances, but also some changes in social and scientific attitudes, before they become feasible. Therefore, we will begin by discussing the social and scientific climate that exists at present concerning searches for extraterrestrial life. Since we are most familiar with the U.S. space program, we will concentrate on it, even though the search for extraterrestrial life is properly the concern of all human beings. Next, we will discuss what might be done within the solar system over the next half century, based on present technology. Finally, we will describe some longer-term projects for exploration both within and outside the solar system.

TWO FAILURES OF NERVE
ABOUT EXTRATERRESTRIAL LIFE

In the 1960's, amidst thé euphoria and ample budgets generated by the Apollo project, Americans looked forward to an ambitious and exciting future in space. Of the various enterprises in space that this country was to undertake, few generated as much enthusiasm as the search for extraterrestrial life. We have disagreed with Professor Norman Horowitz on many other matters elsewhere in this book, but we heartily endorse the words he wrote in 1966, "It is generally agreed that the search for extraterrestrial life is the most important scientific objective of the space program." Although he felt, even at that time, that the chances of finding life on Mars were low, the importance of the discovery made such a project very worthwhile. The search for extraterrestrial life resembled a bet placed in a sweepstakes. The chance of success might be low, but if the payoff were large enough, the entire enterprise would be worthwhile. We feel, of course, that the probability of success is not necessarily small, and so find the search even more attractive.

The same theme has been stressed repeatedly by others. In an introduction to a 1972 symposium concerning plans for the Viking project, a Cornell astrophysicist, Dr. Thomas Gold wrote:

The discovery of life elsewhere in the universe would

surely be one of the greatest discoveries of all time. . . . The National Aeronautics and Space Administration and all of its advisory bodies decided almost unanimously that Mars and the search for life on it should be given the most prominent place in the unmanned space program for the next few years. The Viking program resulted from this decision.

This search was not intended to cover Mars alone, nor was it to be limited to the single Viking project designed to explore that planet. Shortly thereafter, Dr. Richard S. Young of NASA headquarters wrote:

The exploration of Mars will be a sequential study, in which one series of investigations will be designed to follow up on the results of earlier investigations. . . . In a proper search for knowledge concerning extraterrestrial life and the origin of life, NASA must explore as many extraterrestrial bodies as possible in our solar system for the relevant information they can supply.

This enterprising spirit, unfortunately, proved to be very short-lived. By the end of the 1970's, the outlook had changed considerably, and the prospects for any reasonable search for extraterrestrial life were much dimmer. The Viking project had been carried out successfully, although, as we have seen, it left the question of life on Mars unresolved. Of various possible NASA planetary missions for the 1980's, only one had been approved—Project Galileo, a Jupiter orbiter probe. Galileo has a number of worthwhile scientific goals, but few that relate to the search for extraterrestrial life. Even this solitary effort had passed the U.S. Congress with difficulty, as the House Appropriations Committee had recommended that it be killed.

Other actions in Congress also reflected animosity toward the space program. A Senate subcommittee, headed by Senator William Proxmire, moved to deny further funds for research on Moon rocks and for attempts to detect radio signals that might originate from advanced civilizations elsewhere in our galaxy. The projects were facetiously given a "Golden Fleece" award, a distinction reserved for undertakings thought by Proxmire to be particularly wasteful of taxpayers' money. (Perhaps scientists should establish a "Lead Jackass" award, to be given to the politician who demonstrates the greatest ignorance in his statements about science.)

The mood in Washington had clearly changed. In part, this change reflected a wave of fiscal restraint that had taken hold by the late 1970's. Constraints had been applied to many other government agencies and programs in addition to NASA. The mood probably did not reflect any unusual financial crisis in the United States, but was rather one of those periodic cyclic shifts in attitude that have often taken place in our history. If so, a more friendly attitude toward space exploration might be expected to happen at some future time, and the search for extraterrestrial life would resume.

A more serious change in attitude had occurred, however, among the scientists closely connected with our space program. It can be illustrated by considering the plans for further exploration of Mars. Earlier, we discussed the original attitude taken toward the design of the Viking experiments. It was felt that a clear positive result would be spectacular and confirm the presence of life on Mars. A negative set of results, on the other hand, might only indicate that we had selected an inappropriate set of tests. Such results could not even establish the absence of life at the Viking sites, let alone elsewhere on Mars. In fact, the results were neither positive nor negative, but ambiguous and contradictory. If the philosophy of a decade ago was to be honored, then the clarification and extension of the Viking results would merit the highest priority in any further investigations of Mars. This has not been the case.

A Mars Science Working Group, chaired by geologist Thomas Mutch, was formed at the request of NASA and drew up a report in 1977 with recommendations for further exploration in a 1984 mission. The prime strategy it recommended would have done much to increase our understanding of the possibility of life on Mars. It proposed a surface rover with a considerable capacity for chemical analysis and an ability to perform simple metabolic tests. But the format and attitude of the report deemphasized the search for life. It stated, "The odds given by various scientists for the existence of life on Mars today vary considerably, but they are usually small. It is therefore reasonable to assign higher priority in Mars '84 to the exploration of geophysical and geochemical questions . . ." (Note the great change in attitude from the sweepstakes position of Professor Horowitz.) In accord with this philosophy, the page of the report assigned to Science Objectives was concerned with the planetary structure and dynamics, chemistry and mineralogy of rocks and surface materials, geology of landforms, surface processes, atmosphere, and magnetic field of Mars. The search for extraterrestrial life was not cited as a principal goal.

While the above list contained a number of valid scientific objectives, it lacked the inspiration connected with the quest for life beyond Earth. It failed to excite either the Congress or the public at large. The mission was an obvious target for those interested in a cutback on spending, and no funds were voted for it.

The lesson to be drawn from this series of events may unfortunately not have been absorbed by scientists. A debate took place in 1979 at the Second International Colloquim on Mars concerning the strategy for a follow-up mission to Mars. The merits of orbiters, rovers, and sample return missions were discussed. The goals advocated, however, had not altered from those described in the Mutch report. At the end, according to *The New York Times*, the chief scientist at the Jet Propulsion Laboratory of the California Institute of Technology stated, "The fact is, there is no next mission to Mars."

Clearly, the aspirations and plans of many scientists in our space program have diverged sharply from those stated a decade ago and from the interests of the public at large. This situation was clearly demonstrated at a meeting of the American Association for the Advancement of Science held in Washington, D.C., in 1978. A symposium was arranged on the topic "Solar System Exploration: Should It be a National Commitment?" in order to rally support for such a commitment. The following statements on the program blurb described the meeting:

> *We are beginning to understand how atmospheres evolve and, by virtue of studying the dynamics, chemistry, and evolution of the atmospheres of our sister planets, Venus and Mars, we might get an insight as to what the future has in store for our own planet Earth. . . . Our program will ultimately result in an understanding as to what the conditions were that caused life to begin on the Earth, and not on any other planet.*

Thus, any possibility of the discovery of extraterrestrial life was dismissed out of hand. The response of the public to the event was minimal. Although a room seating three hundred had been reserved, the maximum number of people in attendance, including the participants, was sixty.

Later the same day, another AAAS symposium was held on the "Search for Extraterrestrial Intelligence." This had been arranged by a student group called the Forum for the Advancement of Students in Science and Technology. All seats were taken in a room about half the size of that reserved for discussion of "Solar System Exploration." Additional spectators sat in the aisles, stood in the rear, and jammed the doorway and just outside the room. The proceedings were broadcast by satellite to another auditorium in California, where additional hundreds of spectators attended. The interested public had cast a clear mandate for the earlier goals of the space program.

The source of enthusiasm for this type of science is connected to the one responsible for the enormous success of the film *Star Wars*, the television series and film *Star Trek*, and for science fiction literature in general. Much of this enthusiasm is unfortunately diverted to concepts which, when inspected in any depth, appear to be nonsense: visits by astronauts in ancient times, astrological influences, pyramid power, and the like. There are, however, authentic, if speculative, questions in which popular interest and legitimate scientific concerns can coincide. In this book we have tried to emphasize the ones concerning the place of life in the Universe. If these questions are brought back to their proper place in the space program, there will be a much better chance of obtaining additional planetary missions in the near future. On such missions, as on Viking, there would of course be ample opportunity to collect atmospheric and geological data. The amount of such data collected may be less than the specialists in those fields would desire, but it is clearly more than they will receive if no missions at all are funded!

We do not want to argue that public enthusiasm should be the dominant factor in deciding the directions that scientific research should follow. In most cases it is more plausible for scientists to lead public opinion about the aims of scientific research rather than follow it. However, wherever a major unsolved scientific problem coincides with a widespread public interest in an area of exploration, it would be foolish not to take the opportunity to satisfy both interests with one program.

While it is clear that changes in attitude have taken place within the U.S. space program, it is harder to pinpoint the exact reasons for these changes. It is unlikely that the importance to science of the discovery of extraterrestrial life has been devalued. At least, we know of no arguments to this effect that have been presented. The explanation must lie elsewhere. It may well be that the actual nature of the results from the Viking life-detection probes proved traumatic and unnerving to some of those involved. The aspirations and nerve of the life scientists may have been abandoned on the plains of Chryse and Utopia along with the Viking landers. It must certainly be frustrating to work for years on the design and delivery of an experiment that one hopes will be definitive, only to find that the results are ambiguous and cannot be interpreted readily in terms of the original protocol. In that sense the life-detection results did poorly when compared with the clear and unambiguous data delivered by the geological probes. In such circumstances it would be quite natural to become discouraged with life-detection experiments and to prefer firm and secure scientific questions. In our interviews with Viking project experimenters, we found several who wished to withdraw from future life-detection efforts.

Such psychological factors have a legitimate role to play in the choices of an individual scientist. They are not relevant, however, to the larger debate on what our national goals in space exploration should be in the future. We have attempted to construct one side of that debate by collecting statements which, taken together, would represent a coherent argument against a further search for extraterrestrial life in our solar system:

> *Mars, among all the extraterrestrial objects orbiting the Sun, is deemed the most likely to have or to have had living inhabitants.*
>
> *All of the data taken together would seem to point toward nonbiological explanations for* all *of the observed reactions in the Viking experiments.*
>
> *At least those areas on Mars examined by the two spacecraft are not habitats of life.*

From which one can proceed to conclusions like the one already cited by Lewis Thomas, that Mars is stone-dead. We then arrive at the final position that if Mars is dead there is no point in looking for life elsewhere in the solar system.

The above argument, while coherent, can be severely criticized in every assumption. First, while Mars has been a favorite science fiction subject, it is still one of the less likely places in our solar system for life to exist (though it may indeed contain life). There is no indication that Mars, as it exists today, has a suitable medium for the evolution of life. If life of the type searched for does exist on Mars, then either there are hidden liquid areas or life developed earlier when liquid was present on the surface. Second, the nonbiological explanations of the Viking results are as speculative as the biological ones. Uncertainty, rather than dogmatism, is the justified response to the Viking results concerning Martian biology. If the nonbiological explanations of the signals were correct, this fact still would not establish that life was absent. As we stated before, it could be that life was present but unable to respond to these particular tests. Finally, we have no information concerning the presence or absence of life on other sites on Mars. Nothing in the Viking results, therefore, provides a reason why we should discontinue our search within the solar system.

WHAT SHOULD WE DO NEXT?

The political developments of the last few years have been discouraging, but they do not alter the validity of our earlier purpose. Since the discovery of extraterrestrial life forms would be of great scientific and cultural importance, it is time that we returned to the quest. Of course, there is ample scope for discussion on the best strategy to be used. Should a return visit to Mars receive the highest priority? If so, should we try for a rover, a return of soil samples, or a sequence of missions combining both? (We discussed these alternatives in Chapter 9.) If not, what body should replace Mars as the object of our immediate interest: Venus? Titan? Jupiter? Perhaps we should perform close preliminary reconnaissance of all of them before we decide. Each option has its merits, and we don't wish to endorse unequivocally any one of them here. We feel that the advantages and disadvantages of each should be explained to a wider public, who should be invited to participate in the debate in order to arouse interest in the decisions. The policy until now has been to keep strategic considerations a rather private affair limited to NASA and to a very small group of the scientific community.

Whatever the actual order of investigation of the various objects in the solar system, we feel that it is important that certain principles be followed:

1. It should be clearly stated that the search for extra-terrestrial life forms is a primary goal of our explorations.

2. We should acknowledge that this goal is important enough to warrant the investigation of a number of bodies in our solar system. If one or more sites gives negative results (this might not be the case!), this should not be sufficient reason to abandon the search.

3. A continuing series of missions will be needed for each site investigated, as in the Apollo project. We should avoid an approach where final answers are expected on the basis of a single trial.

4. Each environment should be examined with an open mind as to the types of life that might exist there. We should not assume that we will find Earthlife or nothing.

5. In our approach to each planet or satellite, we should first attempt to establish whether there is a chemical and physical basis that is suitable for life rather than try to detect life directly. This means that while the search for life should be a central theme of our extraterrestrial expeditions, we need not always make a biological test the first step in examining an environment. On Venus, for example, we want to know whether liquid of any sort is present on the surface, and ascertain the detailed chemical nature of the clouds. On Mars we want to learn whether organic matter is present on other sites on the planet, and whether there are subterranean pockets of water. The chemical composition of the atmosphere of Jupiter is a question of great interest, in par-

ticular the kinds and amounts of organic and other complex compounds.

It would be very desirable to place a gas chromatograph—mass spectrometer (GC-MS) of the type used on Viking within the atmosphere. (In Project Galileo, a probe will be sent into the atmosphere with a mass spectrometer, but unfortunately no gas chromatograph. We will get very little of the type of information we need from it.) On Titan we want to know the composition and temperature of the atmosphere and the surface, and the identity of any liquids that may be present. The answers to these questions would enable us to infer what types of life are most likely to exist there, and to design intelligent follow-up missions to test for them.

Projects of this type fall within the responsibility of NASA and the corresponding agencies in other countries that control space exploration. However, they are not the only ways we could search for other types of life. We have emphasized that unusual life forms may exist on Earth. A search for them could be carried out independently from those in outer space and at much less expense. Furthermore, looking for alternative life on Earth would come under the responsibility of a different set of agencies, and so would not compete for the same funds as other space projects. We believe that a systematic search of this type is well worth doing, in view of the lack of information in this area and the importance of the results.

There are many things for us to do over the next few decades, but there are even brighter long-term prospects for studying the place of life in the Universe.

LATER STAGES IN SOLAR SYSTEM EXPLORATION

In thinking of the exploration of extraterrestrial environments, we usually consider the most obvious problem to be getting there. It is less than twenty-five years since the first man-made artifact was launched without returning to Earth, and even less since the first probe reached another planet. Yet the problems of transporting reasonable-sized objects to the vicinity of any other body in the solar system have essentially been solved in principle, and given suitable dedication, they would soon be solved in practice. With a combination of advanced propulsion systems such as ion rockets and judicious use of the gravitational forces of the planets, we can send a probe almost anyplace in the solar system. We will not go into the details of these propulsion methods here, but suggest that the interested reader turn to this chapter's reference section for more information concerning these methods.

A search for life in most of the solar system requires not only that we send something there, but that probes (and perhaps people) be able to survive in environments that are highly unfit for human existence. As an example of the problems we face, none of the probes that have landed on the surfaces of Venus thus far have lasted more than a few hours because of the high temperatures and pressures, and perhaps because of corrosive substances found there. In order to explore the plains of Venus, the depths of Jupiter, the seas of Titan, and the Sun itself, there will have to be major advances in our ability to design probes that can function in various un-Earthlike conditions.

Very probably, these probes would have to be unmanned at the outset. The already serious problems in designing equipment that can operate under extreme conditions would be greatly aggravated if human beings must also be able to function there at the same time. This is just another way of saying that an organism fitted for life in one environment will find most other environments unlivable. The development of computer and communication technology over the next century should enable unmanned probes to gather data almost as well as human beings. Furthermore, the time for radio messages to travel back and forth within the solar system is short enough that a versatile unmanned probe could be given instructions to modify its activities on the basis of information already sent back. For example, a probe of Venus that discovers a sea of liquid sulfur could be told what further measurements to make in order to search for life that might have evolved in that sea. Therefore, we will discuss here mainly advanced unmanned probes, although if we do discover extraterrestrial life, the temptation for people to follow the probes will become irresistible.

Of the various problems that confront us in exploring the solar system in depth, probably the most difficult to overcome is high temperature. Other problems, such as extreme cold, corrosive chemicals, and low pressures, would seem to have straightforward solutions that are within the reach of present technology. A vigorous program of exploration of environments where such problems arise would probably find these solutions quickly. It therefore seems well within our capabilities to search for life on any of the bodies that have solid surfaces, including Mercury, Venus, Mars, the various satellites, and the asteroids. It would also be feasible to examine the center of several comets to see what is going on there. We will have somewhat more difficulty looking for any life below the frozen surfaces of Ganymede and other icy satellites, as that search could involve boring holes through many kilometers of ice under conditions of intense cold. But nothing in the laws of physics or the properties of materials would make this impossible, and so eventually we will probably do it.

On the other hand, we have no real idea at present about how to deal with really high temperatures combined with high pressures. All the materials we know melt at temperatures well below 4000° C, so any exploratory probe we can now make would rapidly stop functioning in such places as the deep interior of Jupiter or the surface of the Sun. The fundamental reason for this is that the atoms in a solid body are attached to one another by forces of limited strength. These forces resist impacts of other atoms from outside the solid to an extent, but when the impacts of the outside atoms are too energetic, as happens at high temperatures, the forces in the solid are too weak to hold it together and it tends to melt or boil. If we are to explore regions of high temperature, then instead of using ordinary matter, our probes must use forms of matter that are stable under those conditions. Just as one cannot explore Antarctica wearing clothes designed for the Amazon jungle, so we must adapt our instruments to fit the conditions of the place we want to explore. At present we do not know how to do this, but in principle there is no barrier to making such instruments.

In order to explore the regions of high temperature and pressure which exist deep inside the planets, it may be feasible to utilize the fact that many solids have much higher melting points under high pressure than at Earth's surface pressure. For example, at the pressures deep inside Jupiter, which are millions of times higher than at Earth's surface, ethyl alcohol, the important constituent of vodka, remains solid up to a temperature of 10,000° C. Hydrogen itself is thought to be solid and even metallic at similar temperatures and pressures. It is possible that some of these phases of matter, formed at high pressure, would remain metastable at ordinary pressures. At least one example of this is known. Diamonds are a phase of carbon, formed at high pressure, which, although they do not last forever at ordinary pressures, are stable enough for most purposes. Under laboratory conditions, we may be able to forge instruments made of materials that are stable at high temperatures and pressures, transport them in a metastable condition to the high-temperature environments we wish to explore, and then use them in those environments where they are truly stable.

It is also necessary to avoid destructive chemical and physical reactions between the material in the probes and that of the planetary interior. These reactions will, as we have seen, tend to be speeded up by the high temperatures and pressures. We cannot expect that any material will be immune to all such reactions. But in each of the environments in the solar system, outside of the Sun itself, there are some atomic or molecular forms of matter that are relatively stable physically and chemically. It is these stable forms which would be most suitable for the construction of probes to study each specific environment. A good deal of study, including much trial and error, will be needed before we can do this. The understanding of the properties of materials under conditions very different from those of Earth is one of the main ways in which our science and technology will advance as we explore the solar system.

In the interior of the Sun and other stars, the temperatures and pressures become so high that no matter in atomic form is stable. Therefore, some other form must be used to explore such environments. To do this, we might be able to develop instruments that are constructed along the lines of the plasmobes; that is, out of moving charges which are stabilized by magnetic forces. Such "plasmachines" would operate within a star by using their own magnetic forces to divert the constant bombardment by the particles in the stellar plasma. They would also need to reflect, or otherwise radiate away, the high flux of radiation present inside a star. As we learn to manipulate plasmas more effectively, as part of a program of controlled nuclear fusion, we should determine how feasible it is to construct plasmachines.

It is difficult to estimate how long it will take us to develop the technologies that will enable us to explore the high-temperature regions of the solar system, but it will probably be one or two centuries before we can seek out the forms of life that may be thriving there.

BEYOND THE SOLAR SYSTEM

What of the long-term possibilities of searching for extraterrestrial life, especially beyond the solar system? There are at least two independent approaches to doing this. The one that is easier to do but harder to interpret is to use the results of observations made from Earth or from near Earth to draw conclusions about circumstances elsewhere in the Universe, including the presence of life. We have described in Chapter 9 some past efforts to do this for Mars, and have discussed the ambiguities of the results obtained. Such ambiguities would very likely be at least as serious in any future attempts to learn about extraterrestrial life from Earth, but that does not mean we should not try them.

For example, there have been several recent proposals, some quite elaborate, for the detection of *intelligent* life elsewhere in the Universe. These would attempt to pick up radio or other electromagnetic signals sent out by such hypothetical beings either to announce their existence or just as incidental by-products of their everyday activities. These ideas have been discussed in some detail in a number of excellent books, among which is one by I. Shklovskii and C. Sagan, *Intelligent Life in the Universe*. Since we have not focused on the question of whether there is intelligent extraterrestrial life, we will not discuss the virtues and faults of this particular strategy. But we note that it is very specific for finding life which is capable of one definite technological feat. Furthermore, a similar technique applied to Earth from elsewhere would have given negative results for Earthlife up until the last fifty years; that is, for all but one part in one hundred million of Earthlife's existence. Of course, this specificity may be somewhat balanced by the greater rewards of detecting intelligent life as opposed to just any old type of life.

There are other kinds of long-range observations that could tell us some things about extraterrestrial life. For example, very large telescopes, built outside of the distorting influence of Earth's atmosphere, could detect the presence of planets similar to Earth orbiting some of the nearby stars. It may be more feasible to do this by detecting the infrared radiation or radio waves emitted by a planet than by detecting its visible light, because a planet emits more radio waves and infrared radiation relative to its star than it does visible light. It may also be possible to learn the temperature and atmospheric composition of extrasolar planets from the same measurements, as has already been done for some planets and satellites within the solar system. In this way we might find out about an extrasolar planet that is cold enough for liquid hydrogen to exist on its surface, a possible home for H-bits. Or we might learn about a planet of Earth size, at about 300° K, where carbon-water life could flourish. Observations of this type have not yet been attempted and probably would require new astronomical equipment built especially for the purpose. This approach is nevertheless likely to be much less expensive than the other way of learning about planetary conditions outside the solar system.

The other way of learning about extrasolar life is to send something or someone there to look directly. This is very difficult to do, mainly because of a mismatch among three things: the human life span, the distance between stars in this part of the galaxy, and the maximum speed at which our vehicles can travel. We have seen that stars are typically ten light years apart. For these trips COSMEL will not help us! In the foreseeable future, we will be able to make spaceships that can travel at a few percent of the speed of light. This would be a thousand times faster than our spaceships have traveled so far. But even at such high speeds, a trip to nearby stars would take many centuries. Therefore, some scientists have concluded that interstellar travel is impossible.

This conclusion is very likely unwarranted. We cannot change the distance between stars, and no good way has been suggested to make a spaceship travel at more than a small fraction of the speed of light. But we can extend the time available for space travel in several ways. One method that may become available is to extend the individual human life span considerably. If human beings lived thousands of years instead of seventy years, centuries-long trips would not be unimaginable. While we do not at present know whether people can live that long, it is not out of the question that some day gerontologists will find out how to extend our lives to this extent, or perhaps that biological engineers can create variations on the human species which live for millennia, rivaling

Sequoia trees as the longest-lived form of Earthlife.

Another possible approach to long space trips would be to find a way to "suspend animation" so that human beings could take a trip lasting centuries without any increase in their active life span. Several speculative methods for doing this have been proposed. One is cooling people to temperatures near zero degrees Kelvin; another is by chemically induced hibernation. We cannot at present do any of these successfully, but models for them exist among other forms of Earth, from tardigrades to bears, and it may turn out to be easier to find a biological solution to the problem of interstellar travel than a physical one. With some form of suspended animation, an interstellar spaceship might have a large crew, of which only a small number would be "awake" at any time. The rest would travel suspended and be awakened in small groups to replace the previous crew, who would themselves be suspended until it was time for their next "shift." If a spacecraft had twenty such crews, each going through two ten-year shifts, a four-hundred-year trip could be accomplished within a working life of twenty years of each crew. At a speed 2 percent that of light, this would be enough for a trip to the nearest star and back.

Yet another approach to interstellar travel is to transport a whole society rather than a single ship. In this way a dedicated group of human beings could, over several generations, travel comfortably to the neighborhood of other stars and return, without the need for any major advances in human biological engineering. What is most needed to accomplish this is an advance in social institutions that would make it possible for a group of human beings to live alone together, dedicated to a single goal for several generations.

Probably by the time we are technologically ready to launch such expeditions, there will have been a good deal of experience with such minisocieties. A powerful case has been made by Gerard O'Neill and others for setting up artificial planetoids, orbiting the Earth or Sun, and large enough for thousands of people to inhabit. Such space settlements may be an important new home for human life in the next century. If this development occurs, such space settlements will play a major role in the future exploration of the solar system and beyond, just as the first European settlements in the Western Hemisphere were bases for further exploration of the Americas.

Space settlements orbiting other planets in the solar system would be convenient bases for exploring these planets, much more so than Earth itself. Mobile space settlements could be the vehicles to transport multigeneration expeditions to other stars. The main problem would appear to be a source of energy. The "local" space settlements that O'Neill has proposed would get their energy from sunlight, which is not an intense enough source to propel a space colony on an interstellar trip at any decent speed, especially since the amount of sunlight decreases as the spaceship leaves the solar system. Only some form of nuclear energy appears sufficient to do that. In order to achieve interstellar travel, we will have to use nuclear fusion. Proposals to do that have already been put forward, using nuclear explosions, and it will probably become technologically possible to begin manned interstellar trips within the next two centuries.

But even before this time, we will be able to send unmanned probes to other stars. Computers are not limited by the human life span and can function without the complex life supports that we need. Therefore, a spaceship with only a computer aboard could be much smaller than one with a human crew. The British Interplanetary Society has done a design study for an interstellar rover with only computers aboard that might be launched within the next century. This rover would weigh only about as much as a large ocean liner, and would be able to make a one-way trip to the vicinity of a nearby star in about fifty years, using nuclear fusion as an energy source. It would not be able to do a detailed life search because it would have no way of slowing down, and so it would pass through any system of planets in about one day. It could, however, probably classify the planets that may exist around the target, Barnard's star. Later probes, with the capability of slowing down, would be needed to look further into anything interesting that the first one discovered.

Eventually, if the human race continues along its present path of technological development, the prospects for more elaborate interstellar exploration seem good. Once this is possible, we will be able to look for life in the vicinity of other stars in the same way we will explore our own solar system over the next century or so. The importance of these extrasolar searches depends very much on what we find in the solar system off Earth. If life is common there, it is likely to be common elsewhere as well. We will probably then find many different types of life outside the solar system, including some of those described in earlier chapters, but the general picture of a Universe full of life will already have been set.

On the other hand, if we find no other life than Earthlife in the solar system, then looking for extrasolar life becomes a much more fundamental enterprise. The important questions would then be:

1. Are there planets sufficiently like Earth where Earthlife can and does thrive?

2. In areas outside the solar system which are completely different from anything here, such as the surfaces of neutron stars, has a highly un-Earthlike life evolved?

Both these questions will probably require centuries or millennia to answer, but if we do not find other life in the solar system, we will need answers to them in order to have a clear picture of the scope of life in the Universe.

A FEW FINAL WORDS

Human beings, the conscious branch of Earthlife, have achieved a measure of understanding of how energy and matter behave under various Earthlike conditions. Based on this understanding, scientists have recognized some general laws that govern this behavior. They have then been able to extend these laws to predict what matter and energy would do under very different conditions than those familiar to us. In the case of many nonliving processes, this extension has been successful, and has led to our comprehension of things as diverse as the radiation emitted from artificial atoms and quakes in the crust of neutron stars.

The ideas presented in this book about extraterrestrial life are an attempt to extend this procedure to life, the most interesting and complex expression of energy and matter, and to speculate about what other varieties may be found elsewhere in the Universe. The main problem in doing this is not a lack of understanding of the general properties of matter and energy. It is that we do not know the precise conditions to be found in various places, and that it is hard to imagine what complexities can emerge from general laws in novel circumstances.

Because of these difficulties, it may be that the specific organisms we have suggested in this book for various extraterrestrial environments are different from those that will be found when people eventually search those areas. We would not regard this as a disproof of our general point of view. Our principal thesis is that many of the diverse environments in the Universe contain life, and that radically different types of life develop and survive in these varoius environments.

Humans and the inhabitants of any of the other isolated biospheres that make up a Universe of life cannot be faulted for concluding that their own environment is uniquely suited for life. The reciprocal influence between life and an environment gradually leads to a high degree of mutual fitness between them. It is not easy to see beyond this mutality to other equally fit solutions of the equations of life. Yet, were we gifted with a vision of the whole Universe of life, we would not see it as a desert, sparsely populated with identical plants which can survive only in rare specialized niches. Instead, we would envision something closer to a botanical garden, with countless species, each thriving in its own setting.

If this latter vision is accurate, another argument for the distinction of Earth and its inhabitants will disappear. Our type of life will not be unique in the Universe. In exchange for this loss of distinction, we will gain something perhaps more desirable. Earthlife will be a part of the much more encompassing phenomenon of Universelife. Just as people gradually extend their vision from self-preoccupation as children to identification with a wider community as adults, so in the coming centuries the human species may prefer to be a local representative of Universelife rather than exist almost alone in a vast and sterile void. If so, this shift in attitude may prove to be the most important outcome of a successful search for extraterrestrial life.

The Levels of COSMEL

═══════════════════════════════════════

THE LEVELS OF COSMEL

I: Above Ground Level

Level	Objects, 0.5 to 5 Meters in Size on This Level
26 (not in COSMEL)	the Universe
25 (top level)	
24	
23	cluster of galaxies
22	
21	our galaxy
20	
19	
18	
17	a galactic cloud
16	one light year
15	
14	
13	diameter of Pluto's orbit
12	diameter of Jupiter's orbit
11	diameter of Earth's orbit
10	the star Antares
9	our Sun
8	Jupiter
7	Earth

Level	Object
6	distance, New York to Chicago
5	
4	height of Mt. Everest above sea level, neutron star
3	ten city blocks
2	skyscrapers
1	tall trees
Ground (zero) level	human being, elephant

THE LEVELS OF COSMEL

II: Below Ground Level

Level	Objects, 0.5 to 5 Meters in Size on This Level
Ground (zero) level	human being, elephant
−1	mouse
−2	mosquito
−3	tiny insect, small letter "o" in book
−4	paramecium
−5	human cells
−6	bacterium
−7	virus
−8	ribosome
−9	molecule of table sugar
−10	atoms
−11	
−12	
−13	
−14	nucleus of a large atom
−15	nucleus of a hydrogen atom

References

CHAPTER 1

PAGE

21 A history of human thought about life on other worlds, including the fate of Giordano Bruno and the Whewell-Brewster debate, is given by S. L. Engdahl in *The Planet-Girded Suns*. New York: Atheneum (1974). The full flavor of the debate can best be appreciated if the original works are read: W. Whewell, *The Plurality of Worlds*. Boston: Gould and Lincoln (1854), and D. Brewster, *More Worlds Than One, the Creed of the Philosopher and the Hope of the Christian*. New York: Robert Carter and Brothers (1854).

22 Some of this history is described by A. Lovejoy, *The Great Chain of Being*. New York: Harper & Row, Publishers (1960), p. 114ff.

22 H. G. Wells, *The War of the Worlds*. New York: Berkeley Publishing Company (1964). (This is just one of many editions.)

23 Sir Francis Younghusband, *Life in the Stars*. London: John Murray (1927).

24 The views of Percival Lowell on Mars can be found in *Mars as the Abode of Life*. New York: The Macmillan Co. (1908). A differing view is expressed by Alfred Russell Wallace, *Is Mars Habitable?* London: Macmillan & Co. (1907).

25 The quotation from George Wald may be found in *Life Beyond Earth and the Mind of Man*, ed. R. Berendzen.

Washington, D.C.: NASA Scientific and Technical Information Office (1973), p. 15.

26 M. Hart, "Habitable Planets around Main Sequence Stars," *Icarus,* 37, 351 (1979).

CHAPTER 2

32 An excellent series of drawings indicating what the Universe would look like in various stages of magnification or reduction can be found in K. Boeke, *Cosmic View, The Universe in 40 Jumps.* New York: The John Day Co. (1957).

33 The square-cube law was introduced by Galileo in *Dialogues Concerning Two New Sciences.* New York: Dover Publications (undated), p. 131f.

45 The nature and structure of viroids are discussed in detail by H. J. Gross and coworkers, *Nature,* 203, 187 (1976).

55 A much more detailed description of the function of proteins and nucleic acids within a cell can be found in recent texts on biochemistry; for example, A. L. Lehninger, *Biochemistry,* 2nd ed. New York: Worth Publishers, Inc. (1975), or J. D. Watson, *Molecular Biology of the Gene,* 3rd ed. Menlo Park, California: W. A. Benjamin, Inc. (1976).

69 A more detailed description of atoms and their constituents is presented by G. Feinberg, *What Is the World Made Of?* Garden City: Doubleday and Co. (1977).

CHAPTER 3

76 The various cycles of matter and energy that take place in Earth's biosphere are described in a series of articles in *Scientific American,* Vol. 223, no. 3 (1970).

80 A clear description of the types of energy and the laws governing their transformation is given by H. Thirring, *Energy for Man.* New York: Harper & Row (1962).

87 A description of some of the organic and mineral substances that can be used as a source of chemical energy by simple living things on Earth is contained in "Nutrition," *Encyclopedia Britannica,* 15th ed., Macropaedia, Vol. 13, 1974, p. 401.

87 Data on the methanogens is given by G. E. Fox, et al., *Proceedings of the National Academy of Sciences,* Vol. 74, 4537 (1977).

88 More information on photosynthesis can be found in an article by Govindjee and R. Govindjee, *Scientific American,* Vol. 231, December, 1974, p. 68.

CHAPTER 4

91 For a discussion of the age of the Universe, see S. Weinberg, *The First Three Minutes.* New York: Basic Books (1977).

93 The likely course of events in the development of life on Earth has been described in greater detail in a number of sources; for example, J. W. Schopf, *Chemical Review of Earth and Planetary Science,* 3, 213 (1975), and S. L. Miller and L. E. Orgel, *The Origins of Life on the Earth.* Englewood Cliffs, New Jersey: Prentice-Hall, Inc. (1974).

96 G. G. Simpson, *The Meaning of Evolution.* New Haven: Yale University Press (1967), p. 18.

97 R. Dawkins, *The Selfish Gene.* New York: Oxford University Press (1977), pp. 16–19.

98 The development of photosynthesis, with its resulting release of oxygen into the atmosphere of Earth, has been described by L. Margulis, J. C. G. Walker, and M. Rambler, *Nature,* 264, 620 (1976).

104 The idea of the common ancestor is discussed by Leslie Orgel, *The Origin of Life: Molecules and Natural Selection.* New York: John Wiley (1973), p. 90.

106 More information on the experiments by Professor Spiegelman and his associates can be found in D. R. Mills, R. L. Peterson, and S. Spiegelman, *Proceedings of the National Academy of Sciences,* 58, 217 (1967), and F. R. Kramer, et. al., *Journal of Molecular Biology,* 89, 719 (1974).

108 More data on the paradox concerning whether proteins or nucleic acids originated first can be found in Miller and Orgel, *op. cit.,* p. 153.

109 The idea of coacervate droplets has been advocated by the Russian scientist, A. I. Oparin. For an account of these views, see S. W. Fox and K. Dose, *Molecular Evolution and the Origin of Life,* rev. ed. New York: Marcel Dekker, Inc. (1977), p. 222.

CHAPTER 5

113 A brilliant early summation of ideas on the origin of life was presented by J. B. S. Haldane, *New Biology*, 16, 12 (1954).

113 A detailed history of the concept of spontaneous generation has been written by J. Farley, *The Spontaneous Generation Controversy from Descartes to Oparin*. Baltimore: The Johns Hopkins University Press (1977). The wheat and underwear recipe is described in D. N. Kenyon and G. Steinman, *Biochemical Predestination*. New York: McGraw-Hill (1969), p. 12.

115 The information and quote concerning F. Wohler are taken from L. F. Fieser and M. Fieser, *Organic Chemistry*, 3rd ed. New York: Reinhold Publishing Co. (1956), p. 3.

115 A description of the carbon compounds in carbonaceous chondrites is given by J. G. Lawless and collaborators in *Scientific American*, 226 (June 1972), p. 38. More detailed information on the amino acids isolated can be found in an article by J. G. Lawless and E. Peterson in J. Oró, et. al., eds., *Cosmochemical Evolution and the Origins of Life*, Vol. II. Dordrecht, The Netherlands: D. Reidel Publishing Co. (1974), pp. 4–8.

116 For further information on the organic molecules in interstellar space, see R. H. Gammon, *Chemical and Engineering News*, October 2, 1978, p. 21.

117 For Leslie Orgel's estimate, see *The Origins of Life: Molecules and Natural Selection*. New York: John Wiley (1973), p. 121.

117 Many accounts have been written about the possible nature of the prebiotic soup. For an example, see S. L. Miller and L. E. Orgel, *The Origins of Life on the Earth*. Englewood Cliffs, New Jersey: Prentice-Hall, Inc. (1974).

117 The early experiments by S. L. Miller and H. C. Urey have been summarized in *Science*, 130, 245 (1959).

118 The geology text is R. F. Flint, *The Earth and Its History.* New York: W. W. Norton Co. (1973), p. 119.

118 This quotation is from B. Rensberger, *The New York Times*, section IV, November 13, 1977, p. 18.

120 This calculation was performed by Harold Morowitz in *Energy Flow in Biology.* New York: Academic Press (1968), p. 5.

122 A computer-based approach to calculating the vast numbers of organic structural isomers is given by D. H. Smith, *Journal of Chemical Information and Computer Sciences,* 15, 203 (1975).

125 S. L. Miller has described the circumstances of his first prebiotic experiment in *The Heritage of Copernicus: Theories "Pleasing to the Mind,"* ed. J. Neyman. Cambridge, Massachusetts: The MIT Press (1974), p. 228.

126 The effect of oxygen in prebiotic amino acid synthesis is described by L. E. Orgel, *op. cit.*, p. 129.

127 The problems in this area have been summarized by J. T.-F. Wong and P. M. Bronskill, "Inadequacy of Prebiotic Synthesis as Origin of Proteinaceous Amino Acids," *Journal of Molecular Evolution,* 13, 115 (1979).

129 See N. H. Horowitz, *Accounts of Chemical Research,* 9, 1 (1976).

129 Some relative elemental abundances are tabulated by C. Sagan in the *Encyclopedia Britannica,* 15th ed., Macropaedia, Vol. 10, 1974, p. 900. A list of common elements in various environments is also given in Table 7-3 of our book.

130 F. Hoyle and C. Wickramsinghe have described their theory in *New Scientist*, November 17, 1977, p. 174; and in *Lifecloud.* New York: Harper & Row (1979).

131 This view of the fundamental laws was presented by M. D. Papagiannis in *Abstract I-A-8*, Second ISSOL Meeting and Fifth International Conference on the Origin of Life, Kyoto, Japan, April 5–10, 1977.

132 The anthropic principle is described by P. C. W. Davies, *The Sciences,* 18, 6 (1978).

134 The theory of panspermia and the arguments against it are

discussed by I. S. Shklovskii and C. Sagan in *Intelligent Life in the Universe.* New York: Dell Publishing Co. (1966), pp. 207–212.

134 Directed panspermia was considered by F. H. C. Crick and L. E. Orgel, *Icarus,* 19, 341 (1973).

134 This proposal by T. Gold is cited in Shklovskii and Sagan, *op. cit.,* pp. 211–212.

135 Ideas concerning gradual chemical evolution have been presented by E. D. Hanson, *Quarterly Review of Biology,* 41, 1 (1966), and J. Keosian in *Cosmochemical Evolution and the Origins of Life,* eds. J. Oró, et. al., Vol. 7. Dordrecht, The Netherlands: D. Reidel Publishing Co. (1974), p. 285. See also M. M. Kamshilov, *Evolution of the Biosphere.* U.S.S.R.: MIR Publishers (1977).

137 The appearance of rings and spirals in chemical reactions are described by A. T. Winfree, *Scientific American,* 230, 82 (June 1974).

137 The ideas of M. Eigen and his coworkers are described in *Die Natürwissenschaften,* 65, 341 (1978) and the references cited therein. An introduction to the work of I. Prigogine may be found in *Physics Today,* November, 1972, p. 23, and December, 1972, p. 38. See also H. Haken, *Synergetics,* 2nd ed., Berlin: Springer-Verlag (1978).

CHAPTER 6

142 This behavior of tardigrades is described in an article by J. H. V. Crowe and A. F. Cooper, *Scientific American,* 225, 30 (December 1971).

142 Some negative comments about trying to define life are given by N. W. Pirie, *Perspectives in Biochemistry* II, Cambridge, 1937. See also P. B. Medawar, *The Future of Man.* London: Methuen (1960), p. 108.

142 An excellent summary of types of definitions of life is presented by Carl Sagan in the *Encyclopedia Britannica,* 15th ed., Macropaedia, Vol. 10, 1974, pp. 893–894.

142 For an example of a definition of this type, see J. M. Ford and J. A. Monroe, *Living Systems, Principles and Relationships,* 2nd ed. San Francisco: Canfield Press (1974).

145 The microbiologists are quoted by J. T. Staley in *Journal of Bacteriology*, 95, 1921 (1968).

148 The idea of the biosphere as an organism is presented by L. Margulis and J. Lovelock, *Icarus*, 21, 471 (1974). See also J. Lovelock, *Gaia—a New Look at Life on Earth*. London: Oxford University Press (1979).

152 The mathematical measure of order is the negative logarithm of the probability of occurrence of the collection. For a discussion of this idea applied to physics, see J. D. Fast, *Entropy*. The Netherlands: Philips Technical Library (1968).

157 The precise meaning of free energy and its properties are discussed by G. N. Lewis and M. Randall, *Thermodynamics*, 2nd ed. New York: McGraw-Hill (1961).

159 H. Morowitz, *Energy Flow in Biology*. New York: Academic Press (1968).

159 E. Schrödinger, *What Is Life?* New York: Anchor Books (1956).

160 Some of the properties of liquid helium are discussed by F. London in *Superfluids*, Vol. II. New York: Dover Publications (1964).

163 An introduction to the properties of lasers is given by A. Schawlow in "Optical Masers," *Lasers and Light*. San Francisco: W. H. Freeman and Co. (1969).

165 Such arguments concerning replication are presented by J. D. Watson in *The Molecular Biology of the Gene*, 3rd ed. Menlo Park, California: W. A. Benjamin (1976), pp. 143–145.

169 The process of nucleation in crystal growth is discussed by C. A. Knight in *The Freezing of Supercooled Liquids*. Princeton: D. Van Nostrand and Co. (1967).

174 Eigen's views are described in *Die Natürwissenschaften*, 65, 341 (1978) and in the references cited therein.

178 J. Lovelock, *Nature*, 207, 568 (1965).

CHAPTER 7

180 W. Whewell, *The Plurality of Worlds*. Boston: Gould and Lincoln (1854).

180 See L. J. Henderson, *The Fitness of the Environment*, re-published with an introduction by G. Wald. Boston: Beacon Press (1958).

184 The use of energy by bacteria during their growth phase has been measured by S. Bayne-Jones and R. S. Rhees, *Journal of Bacteriology*, 17, 123 (1939).

185 The energy flow through the biosphere is estimated by H. Morowitz in *Energy Flow in Biology*. New York: Academic Press (1968), p. 70.

186 The microwave background radiation is discussed lucidly by S. Weinberg in *The First Three Minutes*. New York: Basic Books (1977).

187 The intensities of various types of radiation in space are given by C. W. Allen in *Astrophysical Quantities*, 2nd ed. London: Athlone Press (1963), p. 225.

192 A description of heat engines which extract free energy from temperature differences is given by G. M. Lewis and M. Randall in *Thermodynamics*, 2nd ed. New York: McGraw-Hill (1961), Chapter 9.

193 The data in Table 7-2 is taken from Allen, *op. cit.*, pp. 113, 163, and from R. Smoluchowski in *Jupiter*, ed., T. Gehrels. Tucson: University of Arizona Press (1976).

201 The strengths of different types of bonds are discussed by J. D. Watson in *The Molecular Biology of the Gene*, 3rd ed. Menlo Park, California: W. A. Benjamin (1976), Chapter 4.

203 The data in Table 7-3 on elemental abundances is taken from Allen, *op. cit.*, pp. 30–31, and from Morowitz, *op. cit.* L. Thomas, "The World's Biggest Membrane," *Lives of a Cell*. New York: Viking Press (1974).

212 Reaction rates in solutions are discussed by K. J. Laidler in *Reaction Kinetics*, Vol. II. Oxford: Pergamon Press (1963).

213 Over wide ranges of temperature, the logarithm of the reaction rate varies inversely as the absolute temperature. The temperature dependence is discussed by K. J. Laidler, *Reaction Kinetics*, Vol. I. Oxford: Pergamon Press (1963).

214 H. Urey, as quoted by B. Donn, in *Molecules in the Galac-*

tic Environment, eds. M. A. Gordon and L. Snyder. New York: John Wiley (1973), p. 306.

216 The relation between density and size for a noncollapsing cloud was first discussed by J. Jeans. A simple discussion is given by S. Weinberg in *The First Three Minutes.*

218 A discussion of chemical processes in Earth's atmosphere is given by M. J. McEwan and L. F. Phillips, *Chemistry of the Atmosphere*. New York: John Wiley (1975).

CHAPTER 8

225 The poem is taken from Robert Heinlein, *The Green Hills of Earth*. New York: New American Library (1951), p. 125.

225 L. J. Henderson, *The Fitness of the Environment*. New York: Macmillan Co. (1913), and Boston: Beacon Press (1958) paperback.

226 See G. Wald in *Proceedings of the National Academy of Sciences*, 52, 595 (1964).

227 G. Wald in *Cosmochemical Evolution and the Origins of Life,* eds. J. Oró, et. al. Dordrecht, The Netherlands: D. Reidel Publishing Co. (1974), Vol. I, p. 7.

230 L. J. Henderson, *op. cit.*, p. 79.

230 F. Franks, *Chemistry in Britain*, 12, 278 (1976).

231 F. Franks, *ibid.*

231 A number of alternatives to water as media for life are described by V. A. Firsoff, *Life Beyond the Earth*. New York: Basic Books, Inc. (1963), pp. 108–112, 116–128, 142–144.

232 N. H. Horowitz, *Accounts of Chemical Research*, 9, 1 (1976).

233 Alternatives to carbon as a structural material for life are described by G. C. Pimentel, et. al., *Biology and the Exploration of Mars*, Publication 1296, eds. C. S. Pittendrigh, W. Vishniac and J. P. T. Pearman. Washington, D.C.: National Research Council (1966), and by V. A. Firsoff, *op. cit.*, pp. 128–139.

235 N. H. Horowitz, *op. cit.*

238 An example of a small molecule mimicking the effect of

an enzyme can be found in R. Breslow and D. E. McClure, *Journal of the American Chemical Society*, 98, 258 (1976).

239 For a history of our growth of knowledge about DNA, see F. H. Portugal and J. S. Cohen, *A Century of DNA*. Cambridge, Massachusetts: The MIT Press (1977).

239 The first publication of the theory by Watson and Crick was in *Nature*, 171, 737 (1953).

241 A. G. Cairns-Smith and C. J. Davis, *The Encyclopedia of Ignorance*, eds., R. Duncan and M. Weston Smith. New York: Pocket Books (1977), p. 391.

242 A summary of the conditions such as acid, heat, salt, etc., to which living things on Earth have adapted may be found in an article by P. O. Scheie, *Journal of Theoretical Biology*, 28, 315 (1970).

243 The biology of the cold, dry deserts of Antarctica has been described by N. H. Horowitz, R. E. Cameron, and J. S. Hubbard, *Science*, 176, 242 (1972). The existence of microbes within the rocks in Antarctica is described by E. I. Friedman in the abstracts of *Limits of Life*, The Fourth College Park Colloquium on Chemical Evolution, Oct. 18–20, 1978, Laboratory of Chemical Evolution, Dept. of Chemistry, University of Maryland, College Park.

243 References concerning the possible historical role of the methanogens may be found in a summary by J. F. Wilkinson, *Nature*, 271, 707 (1978).

243 The relation between the methanogens and other microorganisms that are adapted to extreme conditions of heat, acid, or salt are discussed by C. R. Woese, et. al., *Journal of Molecular Evolution*, 11, 245 (1978).

244 The quotes are taken from M. Alexander, *Microbial Ecology*. New York: John Wiley (1971), pp. 4, 14.

245 An example of a microorganism whose growth was inhibited by the presence of others, and which needed unconventional culture techniques for its growth has been described by L. E. Casida, Jr., *Applied Microbiology*, 13, 329 (1965).

247 Professor Wald's opinions on ammonia as a solvent for life

may be found in the *Proceedings of the National Academy of Sciences,* 52, 595 (1964).

248 The relevant details about rates of chemical reactions are given by K. J. Laidler in *Reaction Kinetics,* Vols. I and II. Oxford: Pergamon Press (1963).

249 This comment is from V. A. Firsoff, *op. cit.,* p. 112.

251 A review by V. Amarnath and A. D. Brown of the chemical methods used to synthesize nucleic acids may be found in *Chemical Reviews,* 77, 183 (1977). Note particularly the statement about pyridine on page 199: "The coupling step is usually carried out in anhydrous pyridine, which by far seems to be the best solvent; the reaction is very sensitive to moisture and a special apparatus for it has been described."

253 The complexity of silicate chemistry is described in a number of advanced books in chemistry; for example, P. J. Durrant and B. Durrant, *Introduction to Advanced Inorganic Chemistry.* New York: John Wiley (1970).

254 This possibility has been discussed by A. G. Cairns-Smith in *The Life Puzzle.* Edinburgh: Oliver and Boyd (1971).

254 The idea of silicate life at high temperatures is mentioned briefly by V. A. Firsoff, *op. cit.,* p. 144, and by J. B. S. Haldane, *New Biology,* 16, 12 (1954).

CHAPTER 9

258 D. Brewster, *More Worlds Than One, the Creed of the Philosopher and the Hope of the Christian.* New York: Robert Carter and Brothers (1854), pp. 94, 95.

258 Excerpts from the writings of Giovanni Schiaparelli, Percival Lowell, H. G. Wells (*The War of the Worlds*), Edgar Rice Burroughs, Ray Bradbury, and others may be found in *The Book of Mars,* eds. J. Hipolito and W. E. McNelly. London: Futura Publications Limited (1976).

259 Further information on the theories of Percival Lowell can be found in his book, *Mars as the Abode of Life.* New York: The Macmillan Co. (1908).

259 The speculation about vegetation by Trouvelot is reported

in "Life," Carl Sagan, *Encyclopedia Britannica,* 15th ed., Macropaedia, Vol. 10, 1974, p. 908.

258 A description of the effects of the Orson Welles broadcast can be found in *The New York Times,* October 31, 1938, p. 1.

261 The evidence on the nonexistence of the canals is described by C. Sagan and P. Fox, *Icarus,* 25, 602 (1975).

261 An example of the conviction concerning vegetation on Mars is contained in the Earl Nelson, *There Is Life on Mars.* London: T. Werner Laurie, Ltd. (1955).

262 The quote is from an article by Walter Sullivan, *The New York Times,* July 30, 1965, p. 1.

262 The quote concerning Mariner flyby missions appears in *Possibility of Intelligent Life Elsewhere in the Universe (Revised October 1977).* Report prepared for the Committee on Science and Technology, U.S. House of Representatives, Ninety-Fifth Congress, First Session, Washington, D.C.: U.S. Government Printing Office (1977), p. 12.

263 Accounts describing the surface of Mars, as observed by orbiters and landers, are given in articles by B. C. Murray, *Scientific American,* 228, 49 (January 1973); M. H. Carr, 274, 33 (January 1976); and R. E. Arvidson, A. B. Binder, and K. L. Jones, 238, 76 (March 1978).

269 A summary of the Viking geological and biological results concerning Mars can be found in several articles in *Journal of Geophysical Research,* 82, No. 28 (September 30, 1977), and in *Journal of Molecular Evolution,* 14, No. 1 (1979).

273 The information cited concerning water near the surface of Mars was reported by John Noble Wilford, *The New York Times,* June 8, 1979, p. A19.

278 The quote is taken from K. Biemann, et. al., *Journal of Geophysical Research,* 82, 4641 (1977). The GC-MS results are described in this article and in K. Biemann, et. al., *Science* 194, 72 (1976).

278 The steps that led to the selection of the Viking life-detection experiments are described by R. S. Young in *Origins of Life,* 7, 271 (1976).

279 A description of the cancelled experiment is given by W. V. Vishniac and G. A. Welty, *Icarus,* 16, 185 (1972).

279 The quote by R. S. Young is taken from his article in *Origins of Life,* 7, 271 (1976).

282 The comment by Sagan and Lederberg is taken from *Icarus,* 28, 291 (1976).

282 A fascinating description of the day-by-day reactions of scientists to the results of the Viking Life Detection Package is given by H. S. F. Cooper, Jr., *A Search for Life on Mars.* New York: Holt, Rinehart and Winston (1980).

282 The gas exchange experiment results are discussed in V. I. Oyama and B. J. Berdahl, *Journal of Geophysical Research,* 4669 (1977), and the references cited therein.

286 A brief account of Oyama's ideas on the evolution of the Martian surface is given in *Science News,* 111, 276 (1977).

287 The labeled release experiments are described in G. V. Levin and P. A. Straat, *Journal of Geophysical Research,* 82, 4663 (1977), and *Journal of Molecular Evolution,* 14, 167, 185 (1979).

290 This opinion by H. P. Klein is taken from *Icarus,* 34, 666 (1978). This article also summarizes the chemical explanation of the Viking life-detection experiments.

291 Information of the carbon assimilation experiment can be found in N. H. Horowitz, G. L. Hobby, and J. S. Hubbard, *Journal of Geophysical Research,* 82, 4659 (1977), and J. S. Hubbard, *Journal of Molecular Evolution,* 14, 211 (1979).

296 See Norman H. Horowitz, *Scientific American,* 237, 52 (November 1977).

296 The quote from Lewis Thomas is taken from a column opposite the editorial page in *The New York Times,* July 2, 1978.

297 Jastrow's affirmative conclusion about life on Mars is given in his book, *Until the Sun Dies.* New York: W. W. Norton and Co. (1977), p. 154ff.

298 This quote comes from an article by Carl Sagan in *The New York Times,* February 22, 1975, p. 27.

298 The details concerning color changes on the rocks on Mars have been published by G. V. Levin, P. A. Straat, and W. D. Benton, *Journal of Theoretical Biology,* 75, 381 (1978).

299 These terms were introduced by Sagan and Lederberg in an article in *Icarus,* 28, 291 (1976).

300 B. C. Clark, *Origins of Life,* 9, 241 (1979).

302 Strategies for future Mars exploration are discussed in *A Mars 1984 Mission, Report of the Mars Science Working Group,* TM-78419, National Aeronautics and Space Administration (July 1977); and *Post-Viking Biological Investigation of Mars,* National Research Council, Washington, D.C., National Academy of Sciences, 1977.

CHAPTER 10

311 The data in Tables 10-1 and 10-2 is taken from C. W. Allen, *Astrophysical Quantities,* 2nd ed. London: Athlone Press (1963), Chapter 7. Some updating has been done.

317 The analysis of lunar soil is described in articles by C. Ponnamperuma and others, J. Oró and others, and V. Oyama and others in *Science,* 167, 3918 (1970).

319 Much of the information on Jupiter described here is given in *Jupiter,* ed. T. Gehrels. Tucson: University of Arizona Press (1976).

325 Some interesting speculations on life on Jupiter are presented by E. Salpeter and C. Sagan, *Astrophysical Journal Supplement,* 32, 737 (1976). These authors discuss a variety of topics from physical processes in the atmosphere through the ecology of the living things that may be found there.

327 The properties of liquids and solids at high pressure are discussed by P. W. Bridgeman, *Journal of Chemical Physics,* 3, 597 (1935).

329 The presence of an internal sulfur ocean on Io has been suggested by B. A. Smith and coworkers, *Nature,* 280, 738 (1979).

329 The model of Ganymede that we describe here is discussed by C. J. Consolmagno and J. S. Lewis in *Jupiter,* ed. Gehrels.

332 Life in the vicinity of hot vents on the sea bottom has been described in *National Geographic*, 156, 687 (November 1979).

332 The atmosphere of Titan is discussed by H. E. Hunt, *Advances in Physics*, 25, 455 (1976), and by S. K. Atreya and others, *Science*, 201, 611 (1978).

335 A survey of what is known about Venus is given by D. M. Hunten and others, *Space Science Reviews*, 2, 265 (1977).

337 The hypothetical "airbags" of Venus were described by H. Morowitz and C. Sagan, *Nature*, 216, 5107 (1967).

343 S. Dole and I. Asimov, *Planets for Man*. New York: Random House (1964).

CHAPTER 11

353 A description of the conditions in the center of the galaxy, based on measurements of infrared and other radiation, is given by T. R. Geballe in "The Central Parsec of the Galaxy," *Scientific American*, 241, 60 (July 1979).

355 The description of the interior of the Sun given here is based on D. Menzel and others, *Stellar Interiors*. London: Chapman and Hall (1963).

360 The list of nearest stars is taken from a longer list given by C. W. Allen, *Astrophysical Quantities*, 2nd ed. London: Athlone Press (1963), pp. 225–227.

362 The quotation is taken from J. Greenstein in *Low Luminosity Stars*, ed. S. Kumar. New York: Gordon and Breach (1969), p. xvi.

363 The existence of molecules in red giant stars is discussed by G. Wallerstein, "Molecules in Red Giant Atmospheres," *Molecules in the Galactic Environment*, eds. M. A. Gordon and L. Snyder. New York: John Wiley (1973).

368 A summary of properties that neutron stars are expected to have is given by G. Baym and C. Pethick, *Annual Reviews of Nuclear Sciences*, 25, 27 (1973).

370 The properties of magnetic atoms and chains are discussed by M. Ruderman in *Physics of Dense Matter*, ed. C. Hansen.

Dordrecht, The Netherlands: D. Reidel Publishing Co. (1974).

CHAPTER 12

379 The account by Sir David Brewster, including the comments of Elliot and Herschel, is taken from Brewster's book *More Worlds Than One, the Creed of the Philosopher and the Hope of the Christian*. New York: Robert Carter and Brothers (1854).

380 One author's vision of life in the Sun is given by O. Stapledon in "The Flames," reprinted in *Worlds of Wonder*. Los Angeles: Fantasy Publishing Co. (1949).

380 The influences of magnetic forces on conditions inside a star like the Sun are described by H. Alfven, *Cosmical Electrodynamics*. London: Oxford University Press (1959).

381 Maude's ideas are discussed in his article "Life in the Sun," *The Scientist Speculates*, ed. I. J. Good. New York: Basic Books, Inc. (1963).

381 Some of the mutual effects of charges and magnetic forces in a plasma are described by L. Spitzer in *The Physics of Fully Ionized Gases*. New York: Wiley Interscience (1956).

382 The conditions of radiant energy inside the Sun are discussed by D. Menzel and others in *Stellar Interiors*. London: Chapman and Hill (1963).

389 The best summary of properties of various forms of hydrogen is still A. Farkas, *Orthohydrogen, Parahydrogen, and Heavy Hydrogen*. Cambridge: Cambridge University Press (1935).

395 The distinctive properties of the radiation from lasers are described in an article by A. Schawlow, "Laser Light," *Lasers and Light*. San Francisco: W. H. Freeman and Co. (1969).

396 Some of the information about masers in interstellar space is given by A. H. Cook in *Celestial Masers*. Cambridge: Cambridge University Press (1977).

CHAPTER 13

401 Some of the speculations in science fiction about extrater-

restrial creatures are summarized by J. White in *The Visual Encyclopedia of Science Fiction.* New York: Harmony Books (1977), p. 99.

402 The organization of Earthlife as a system is considered extensively by J. G. Miller, *Living Systems.* New York: McGraw-Hill (1978). Comparisons are made with other systems, such as social organizations.

408 The basis of this exhibit is described by B. Dalzell in "Exotic Bestiary for Vicarious Space Voyagers," *Smithsonian Magazine,* 5, 84 (October 1974).

409 For the quote by MacGowan and Ordway, see *Intelligence in the Universe.* Englewood Cliffs, New Jersey: Prentice-Hall (1966), p. 240.

410 F. Jacob, *Science,* 196, 1161 (1977).

410 G. G. Simpson, *Science,* 143, 769 (1964).

411 Convergent evolution is discussed by G. G. Simpson in *The Meaning of Evolution.* New Haven: Yale University Press (1949), p. 181ff.

CHAPTER 14

416 The quote by Professor Horowitz is taken from an article in "The Search for Extraterrestrial Life," ed. J. J. Hanrahan, Advances in the Astronautical Sciences, Scholarly Publications, American Astronautical Society, Sun Valley, Calif., Vol. 22, 1966.

416 See T. Gold, *Icarus,* 16, ii (1972).

417 The statements by R. S. Young are taken from his article in *Molecular Evolution, Prebiological and Biological,* eds. D. L. Rolfing and A. I. Oparin. New York: Plenum Press (1972).

417 The action of the subcommittee is described in articles by Thomas H. Jukes, *Nature,* 275, 584 (1978), and by David Dickson, *Nature,* 274, 522 (1978).

418 See *A Mars 1984 Mission, Report of the Mars Science Working Group,* TM-78419, National Aeronautics and Space Administration, July 1977.

419 This quote is from an article by John Noble Wilford, *The New York Times*, January 18, 1979.

420 The quote is from "Solar System Exploration: Should It Be a National Commitment?" *Abstracts of the American Association for the Advancement of Science, Annual Meeting*, Washington, D.C., February 12–17, 1977, p. 99.

422 The quote starting "Mars, among all . . ." is found in *Post-Viking Biological Investigations of Mars*, report by the National Research Council, National Academy of Sciences, Washington, D.C., 1977.

422 The conclusion concerning "All of the data . . ." was expressed by H. P. Klein in a summary of the Viking Biological Experiments, *Icarus*, 34, 666 (1978).

422 The quote starting "At least these areas . . ." is from N. H. Horowitz, *Scientific American*, 237, 52 (November 1977).

426 The use of ion rockets for future space missions is discussed by E. Stuhlinger in *Ion Propulsion for Space Flight*. New York: McGraw-Hill (1964).

428 The properties of matter at high temperature and pressure are summarized by S. A. Babb, Jr., in *Reviews of Modern Physics*, 35, 400 (1963).

430 I. Shklovskii and C. Sagan, *Intelligent Life in the Universe*. San Francisco: Holden-Day (1966).

432 These ideas on space colonies are presented by G. O'Neill in *The High Frontier*. New York: William Morrow and Co., Inc. (1977).

433 The proposed Daedalus mission to Barnard's star is described by N. Calder in *Spaceships of the Mind*. London: British Broadcasting Company (1978).

Index

search for life, 417–419, 423–424
temperature of, 24, 259, 269
vegetation on, 259, 261, 278
volcanoes, 262, 270, 273
See also Intelligent life, Lichens, Oxygen, Viking project
Mars Science Working Group, 418–419
Mass spectrometer, 276, 285, 288
See also GC-MS
Maude, A. D., 381
Medawar, J. S., 139
Medawar, P. B., 139
Mendel, Gregor, 102
Mercury (metal), 231
Mercury (planet), 249, 311–312, 338
interior of, 338
Metals, 232, 254
Meteorites, 130, 134, 277
carbonaceous chondrites, 115–116, 130, 275
Murchison, 116
Methanogens, 74, 111, 243–244, 252
Methane, 87, 94, 111, 120–121, 232, 250
Miescher, Friedrich, 239
Miller, Stanley, 117, 125–126, 130
Membrane, cell, 39–40, 43, 55, 96, 104, 108, 237, 242
Mirror image forms of molecules, 52, 55, 104, 116, 284, 288
Mitochondria, 42–44, 402
Mobility of atoms, as requirement for life, 211, 218–219
Molecules, 46, 48–49, 83–85, 189, 201
in red supergiant atmospheres, 363, 374
See also Carbon
Moon, 23, 210, 258, 312–313, 317–318
absence of water on, 228
apparent absence of life on, 414
Morowitz, Harold, 159, 337
Multicelled organisms, 93, 100
Mutation, 101–103, 108, 410
Mutch, Thomas, 418

NASA (National Aeronautics and Space Administration), 256, 261–262, 278, 290, 303, 417–418, 424–425
National Academy of Sciences, 303
National Air and Space Museum, 409
Natural selection, 101, 103, 105, 107, 135, 410
Neptune, 311, 334
Neutrinos, 71, 182, 368
Neutron stars, 71, 206, 219, 368–373, 411
as possible homes for life, 377
composition, 370–373, 375
crust, 370, 375
energy flow onto surface, 373–375

interior, 373–375
size, 368, 370
surfaces, 368, 370–371, 373, 377
Neutrons, 70–71
Nitrogen, 120, 247, 285
chemical properties of, 233, 249
in organic compounds, 53–54, 57, 115, 227, 233
Nonequilibrium, as necessity for life, 197
Nuclear fusion, as energy source, 358, 385–386
Nucleation, 169–170
Nucleic acids
alternatives to, as basis for life, 237, 240–241, 245
as replicator, 108–109, 119–121
chemical synthesis, 251, 449
circular, 68
discovery of, 239
DNA, 55, 62–64, 66–67, 165, 239–240, 242, 404
damage by water, 230–231
on Mars, 286
viral, 67–68
in early life, 96
in prebiotic synthesis, 123, 128, 251
L-shaped (transfer), 61, 65
messenger, 63–66
mutation and, 103
RNA, 55, 65, 67, 106–109
role in Earthlife, 24, 43, 47, 104, 143–144, 242
structure of, 60, 233–234
Nucleoside triphosphate, 65–67, 106, 109
Nucleotide, 60, 64–65, 123–124, 127–128
Nucleus, atomic, 70–71
Nucleus, cell, 39–41, 100

Ockham's razor, 297
O'Neill, Gerard, 432–433
Oparin, A. I., 441
Order, 150–171, 197–200, 204, 401
chemical, 200
creation of, 162–163, 165–169, 171, 198
in collections, 151–153
in magnetic atom polymers, 371
in magnets, 160–161, 163
in non-living systems, 159–161, 163–164, 166, 168
in nucleic acids, 154–155, 157
in plasmas, 381–382
in superfluids, 160, 373
in the biosphere, 153–155
in the Universe, 413
maintenance of, 155–159
measure of, 445
of chemical specificity, 153–155
of polymer sequences, 154–155

461

464